STEPPES

STEPPES

The plants and ecology of the world's semi-arid regions

DENVER BOTANIC GARDENS
Michael Bone, Dan Johnson,
Panayoti Kelaidis, Mike Kintgen,
and Larry G. Vickerman

TIMBER PRESS
Portland, Oregon

Dedicated to Brian Vogt
CEO, Denver Botanic Gardens (2007–present)

Onward!

The past may dictate who we are,
but we get to determine what we become.
—Unknown

Published in 2015 by Timber Press, Inc.

Unless otherwise credited, all photographs are by the authors.

The Haseltine Building
133 S.W. Second Avenue, Suite 450
Portland, Oregon 97204-3527
timberpress.com

Printed in China
Cover and text design by Patrick Barber
Maps by Benchmark Maps
ISBN-13: 978-1-60469-465-9

Library of Congress Cataloging-in-Publication Data

Bone, Michael, author.
   Steppes: the plants and ecology of the world's semi-arid regions/
Denver Botanic Gardens, Michael Bone, Dan Johnson, Panayoti
Kelaidis, Mike Kintgen, and Larry G. Vickerman.—First edition.
      pages cm
   Other title: Plants and ecology of the world's semi-arid regions
   Includes bibliographical references and index.
   ISBN 978-1-60469-465-9
   1. Steppes. 2. Steppe plants. 3. Steppe ecology. 4. Arid regions.
I. Title. II. Title: Plants and ecology of the world's semi-arid regions.
   GB571.B66 2015
   577.4'4—dc23
                                    2014048499

A catalog record of this book is also available from the British Library.

CONTENTS

FOREWORD

WHEN WE TALK about the prairie, our minds conjure up images of homesteaders traveling across the grasslands in wagons. On a global scale, we can visualize the nomads of Mongolia moving from one settlement to another through the vast steppes of Central Asia. The North American shortgrass prairie and Great Plains are one of four major steppe regions of the world, the others being the Patagonian steppe in South America, the Karoo and savanna regions of South Africa, and the Eurasian steppe, which includes easternmost Europe and Central Asia. This last, the largest of the steppe regions, supported important travel and trade routes between Asia and Europe; some portions of the most famous of these—the Silk Road connecting China, India, and Europe, established around 200 BC—are still in use.

A passenger traveling through western Kansas and eastern Colorado might consider the trip boring as she passes through mile after mile of what looks like parched grassland. But if she were to alight from the vehicle, walk through the prairie, and look down, the diversity of life would catch her by surprise. Grasslands of the world sustain an enormous amount of native biodiversity. In addition to grasses and the herbivores that depend upon them, steppe ecosystems harbor numerous species of forbs and shrubs; they are mostly without trees, except for those near lakes and rivers. Even though steppe is dominated by grasses and grass-like species, the number of grass species is frequently lower than forb species.

Plants from steppe regions are adapted to harsh climatic conditions, tolerating extreme heat and cold, high winds, and low moisture. The steppes themselves, located between temperate forests and deserts, are characterized by a semi-arid and continental climate. They generally receive 10 to 20 inches of rainfall annually and experience extreme cold in the winter (−40°C) and high summer temperatures of up to 40°C. In addition to these seasonal temperature swings, they are also subjected to huge differentials between day and night temperatures. These extreme climate conditions have defined and shaped steppe flora and fauna.

Sadly, the steppe regions of the world are threatened by loss of habitat due to human-induced pressures. In regions with higher rainfall, plants produce large amounts of below-ground growth (roots, rhizomes), which translates to high amounts of organic matter in the soils; these enriched soils were (and are) seen as prime sites for agricultural development—but where once the grasses held soil moisture during drought and trapped soil, preventing wind erosion, conversion to croplands has led to severe losses of topsoil (witness the 1930s Dust Bowl in North America). For another example: the steppes can withstand moderate grazing and indeed support a vast array of native herbivores, but they have been greatly impacted by domestic livestock introduced by human societies. Overgrazing causes changes to plant species composition, which impact animal populations. Commonly, this results in conversion of grasslands to shrublands dominated by less palatable herbaceous and woody species, with a consequent loss of biodiversity in steppe ecosystems.

Denver Botanic Gardens serves as an international center for the study of steppe flora. Through our research and trials, plants from the world's steppe regions are tested for their adaptability to Colorado's semi-arid steppe climate. Those with unique and sustainable landscape characteristics are selected and introduced to the public through the collaborative Plant Select program. In the 21st century, as we battle with issues of climate change and reduced water resources, it is imperative that we utilize plants long-adapted to extreme climatic fluctuations. This book provides knowledge about the world's steppe regions and their flora and insight into climate-adapted landscapes. With it, we hope to spark an interest in the steppe ecosystems of the world, create a better understanding of their biodiversity and ecological complexity, and help develop stewardship in conserving these important ecosystems.

SARADA KRISHNAN, PH.D.
Director of Horticulture and the Center for Global Initiatives
Denver Botanic Gardens

INTRODUCTION

PRINCIPAL STEPPE REGIONS

PANAYOTI KELAIDIS

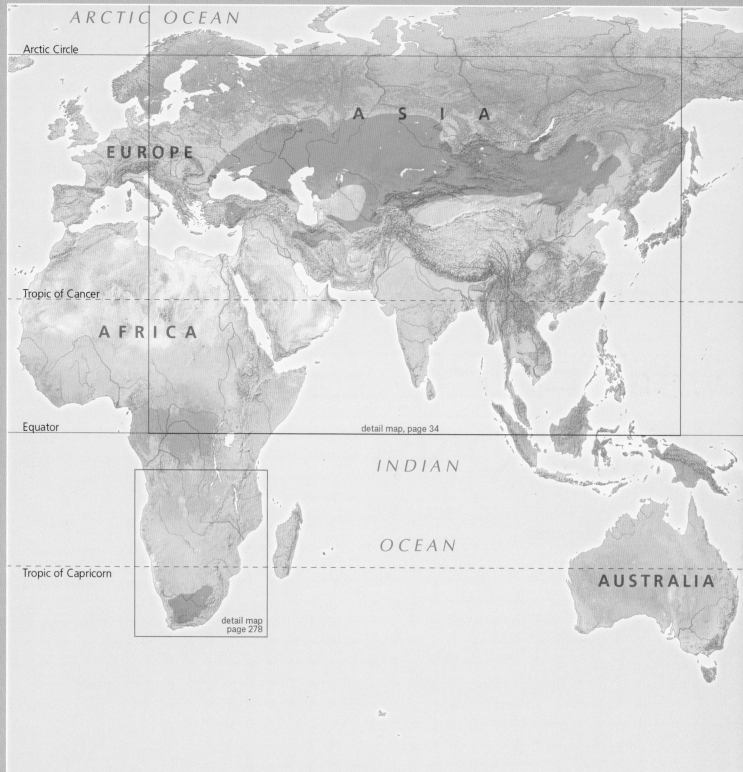

ARCTIC OCEAN

Arctic Circle

ASIA

EUROPE

AFRICA

Tropic of Cancer

Equator

detail map, page 34

INDIAN

OCEAN

AUSTRALIA

Tropic of Capricorn

detail map
page 278

SOUTHERN OCEAN

Antarctic Circle

ANTARCTICA

WORLD STEPPES

ARCTIC OCEAN

Arctic Circle

NORTH PACIFIC

OCEAN

NORTH
AMERICA

detail map
page 138

detail map
page 88

NORTH
ATLANTIC
OCEAN

Tropic of Cancer

Equator

SOUTH
AMERICA

Tropic of Capricorn

SOUTH PACIFIC

OCEAN

SOUTH
ATLANTIC
OCEAN

detail map
page 220

SOUTHERN OCEAN

Antarctic Circle

| 0 | 1,000 | 2,000 | 3,000 | 4,000 | 5,000 kilometers |
| 0 | 1,000 | 2,000 | 3,000 | 4,000 | 5,000 miles |

STEPPES HAVE THE ALLURE of the distant and exotic. They are associated in the popular mind (when associated at all) with the steppes of Russia: in the 19th century, when all things Russian were influencing European thought and culture (and vice versa), the Russian "степь" was transliterated as "steppe," a word corresponding to our "prairie." In Russia, the word was first utilized to describe the grasslands of west-central Russia proper (and neighboring countries once part of the Soviet Union, such as Ukraine and Moldova), extending from the region of Moldavia in eastern Europe at its westernmost extension, in a broad band around the Black Sea and Caucasus. This region in Russia is very similar climatically and culturally to the prairie and plains regions in North America; indeed, steppe supports the bulk of grain production in both continents.

This original grassy version of steppe has accrued a quantity of associations. Many classic works of 19th-century Russian literature include the word in their titles—Maxim Gorki's novella *On the Steppes*, Ivan Turgenev's short novel *King Lear of the Steppes*, Anton Chekhov's haunting short story "The Steppe." Steppe manifests itself in German literature in Hermann Hesse's novel *Steppenwolf* (in turn adopted as the name of a hugely popular 20th-century rock band). Nikolai Borodin's haunting *In the Steppes of Central Asia*, a staple of concert halls and classical music radio, suggests caravans crossing an obscure landscape thousands of miles east of the European grasslands of the Cossacks. In fact, grassland ecosystems fall in a broad latitudinal band that encompasses much of the southern part of the former Soviet Union, between true desert to the south and moister taiga in Siberia proper; and the term "steppe" has come to be applied to a vast swath of Asia—and consequently a wider and more complicated range of climatic extremes.

No climatic region on earth has been more central to the history and peregrinations of humanity. Central Asia was an enormous theater of human prehistory and historical drama: hundreds of khanates and kingdoms arose and vanished on the steppes. Most inhabitants of the region were nomads who left scant trace of their passage. But many steppe tribes settled down, creating cities like Bukhara, Samarkand, and Taraz, leaving behind monuments and archaeological evidence of great artistic interest. In recent decades—thanks mostly to oil and mineral development and general population pressures—humanity is once again invading the steppes. On ecological maps of the world, the steppe biome occupies conspicuous swaths on four continents. And yet it could be argued that steppe—the ecosystem that gave rise to and shaped humanity—remains among the least studied or understood ecosystems on our planet. There has never been a comparative study of the four major steppe ecosystems that constitute the very heart of the four largest continents. This book hopes to fill that gap.

Coming to Terms

What is steppe? There is no easy answer. The term "steppe" is used descriptively for most any temperate grassland or semi-arid shrubland, and by many geographers with specific reference to distinctive ecosystems, throughout the continental portions of both northern and southern hemispheres. "Steppe" has been used as a sort of shorthand to characterize cultural, ecological, edaphic, and phytogeographic phenomena.

Steppe is first and foremost a function of climate, but steppe climates, on all four continents, are a paradox: unlike Mediterranean biomes, for instance, which are often simplistically characterized (wet winters; hot, dry summers), each of the world's steppe regions constitutes a cline rather than a static state between neighboring biomes. At one end, each steppe region has winter precipitation (like the Mediterranean), while at the other end, rain is concentrated in the summer months. Distinctions between and within the various types of steppe are blurry, constantly shifting over the short term. The Venn diagram here illustrates how steppe overlaps with and relates to neighboring biomes. The border separating steppe from Mediterranean biomes may be stark and abrupt (the Sierra-Cascade crest or the Himalaya, for instance), but the transition between steppe and the other two biomes is often less distinct. In wet years, steppe can approximate the rainfall and humidity of a maritime climate, but in drought cycles the same region may have precipitation levels that approach the aridity of true desert—occasionally for prolonged spells. Variable and extreme weather patterns are characteristic of steppes.

The terms used to describe steppe-like environments in the four continents

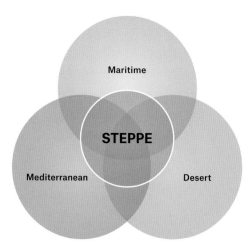

Most steppes represent the transition between maritime climates to the east and more arid or Mediterranean climates to the south and west.

are legion and often confusing, whether employed as vernacular references or in an attempt at greater rigor: prairie, plain, Highveld, matorral, range, Karoo, puna are just some of the better-known labels. Few have had the privilege of exploring all these regions, as we, the authors of this book, have; and most don't realize the extent to which the four most extensive semi-arid continental biomes are similar to one another in terms of land forms, general geomorphology, plant-animal interactions (the nature of human and livestock pressures), and above all, floral makeup. The physiology of landscape and ecological interplay of plant communities within each of these regions are uncannily similar.

In North America, the Middle English word "plains" was used to describe level grassland from the very beginning of English settlement. By 1773, the French word "prairie" was also being used, and in 1823, explorer Edwin James famously labeled the Great Plains "The Great American Desert." But when travelers of European ancestry finally grew familiar with the true deserts, to the west of the Rockies, the linguistic challenge arose: how to characterize the slightly less arid regions of the Great Plains, which more and more pioneers had to cross? By the middle of the 19th century, many of these transients actually began to settle much of the more humid parts of the shortgrass prairie, and the taller grasslands to the east of the 100th meridian had virtually vanished, replaced almost entirely by farms by century's end.

"Steppe" was first used outside Russia by Alexander von Humboldt, after his 1829 visit to the Russian steppes. It is most commonly used in English with reference to the original "steppes of Central Asia," and of course easternmost Europe: it is a Russian word, after all, and "steppe" gained currency in the English lexicon throughout the 19th century. Ironically and approximately concurrently, the German (largely Mennonite) minority in Russia began to be conscripted under Alexander III, resulting in the enormous migration of Volga Germans to the Great Plains of Canada and the United States. Not only did a large population of people move from the steppes of Russia to their steppe equivalent in west-central North America, but they brought with them the red durum wheats of our continent's "bread basket." Intermixed with the wheat seed (which in those days was uncertified and could be imported with impunity) came the entire suite of western Asian weeds that still constitute the bulk of invasive species throughout the West including, for instance, tumbleweed (Russian thistle, *Salsola tragus*), which was introduced accidentally in 1873 in flax seed imported by Russian Germans in South Dakota. Just four families of plants (Asteraceae, Amaranthaceae, Poaceae, and Brassicaceae) constitute the overwhelming bulk of invasive species from western Asia to the equivalent steppe climates of North America.

That the concept of semi-arid, continental climates is relatively new to the English language—since there is nothing continental or semi-arid about the

British Isles—helps explain why steppe in a global context is so poorly understood in North America. Much of the interior American West was settled by feverish miners and pioneers in Conestoga wagons from much more humid climates. These harried settlers—and the American Indians they overwhelmed—were not students of world biogeography. Few 18th- and 19th-century Americans would ever have traveled to Asia, let alone the southern hemisphere correlative climates, to observe the similarities firsthand. As far as most settlers were concerned, much of the West was just plain desert after all.

An even more compelling explanation for why awareness of steppe climate and ecosystems has been delayed as an arena for serious inquiry may boil down to the most prosaic of truisms. Humans may have evolved on the steppes, but by the time the scientific revolution took place three centuries ago, the bulk of mankind (including scientists) lived in humid climates; moreover, most great cities were situated on coasts or navigable rivers. The great universities in North America, Europe, and elsewhere were naturally in these humid cities. The great breakthroughs in early biology—Humboldt in South America, Darwin on the *Beagle*—were conducted by scientists from maritime climates, often on ships. Science as we know it was invented by people who viewed the planet through a misty—or even foggy, shall we say—lens. Ironically, the flora of England (or coastal New England, for that matter), however lush it may appear, is comparatively impoverished compared to many steppe climates in acre-to-acre biodiversity. England is believed to have had only 1000 species of indigenous vascular plants prior to the advent of humans in an area twice the size of Colorado, which in turn has over 3000 indigenous species (and counting): six times the species numbers per unit area. Early scientists who first explored the riches of the steppe can hardly be blamed for being somewhat overwhelmed and puzzled by it.

Comparative Biogeography of Floristic Provinces

Botanists, beginning with Linnaeus himself, noted many genera shared between East Asia and southeastern America. It was only in 1846 that Asa Gray published the first detailed comparison between the floras of East Asia and eastern North America. Many botanists had noted these similarities in the century before, but Gray's was the first extensive and systematic paper to compare two far-removed geographic floras. It can fairly be said to have given rise to the whole field of comparative biogeography. That paper opened the veritable floodgates of research in this arena: in the nearly 170 years since, dozens of books, countless articles, international symposia—all continued the comparative study of these two maritime regions.

Likewise there has been a long and rich history of comparative geographical and floristic study of the world's Mediterranean floras. The five Mediterranean floristic provinces—California, Chile, South Africa, Australia, and the Mediterranean littoral proper—are discrete enough in area, and all five regions are peopled with university scholars: the result is that the striking parallels between the floras of these five regions have inspired similarly voluminous accumulated research.

Botanists have also compared equatorial rainforests of the western hemisphere (Central and South America) and of the eastern hemisphere (Africa and Southeast Asia). Of all ecosystems on earth, it's likeliest that these forests, with their enormous biodiversity, have evolved most directly and consistently from a common ancient ancestral flora in the Mesozoic era, so the commonalities and differences in these different biomes have great significance. Large monographs (e.g., Goldblatt 1993) seek to determine how much the respective floras derive from a common Gondwanaland origin, superimposing cladistic analysis of plant families onto a theoretical reconstruction of climate, geography, and plate tectonics. The universal distribution of bromeliads in the New World tropics and their complete absence in the Old World tropics is the universal example, demonstrating that the bromeliads had to have arisen subsequent to the breakup of South America and Africa.

Botanists have also done some comparative study of the true deserts of the world: as with bromeliads, the prevalence of Cactaceae in the New World and Aizoaceae in the Old provide a gauge for the age of these families, perhaps. The convergence of morphological features in disparate regions in the respective deserts is often used as an example of how climate influences, even shapes, the morphology of xerophytic succulents.

The first inklings of the potential parallels between North American and Asian continental floras may be traced to observations made by Joseph Dalton Hooker after his trip to Colorado with Asa Gray in the summer of 1877. Their observations were published in 1880 as "The Vegetation of the Rocky Mountain Region and a Comparison with that of other Parts of the World," which article ultimately inspired William A. Weber (2003). Weber's study focuses almost primarily on the montane and alpine elements of the Eurasian Altai and North American Rockies, with little reference to the semi-arid or steppe elements of the two floras. These two far-flung works are the only extensive comparisons in the literature between the Asiatic and Rocky Mountain floras.

When compared to some of the world's other major phytogeographic regions, the parallels and commonalities between the various steppe climates are as significant and apparent as the more often studied temperate maritime floras, the equatorial rainforests, the five Mediterranean provinces, or the true deserts. This book is an attempt to bring a greater awareness of the many striking parallels between the semi-arid grasslands and more arid steppes on the four largest continents.

The most obvious parallel between the four principal and widely separated steppe floras is that the very same plant families in very similar proportions are present in each: the grass family first and foremost, followed by the daisy family and the bean family, with a strong presence of another dozen or so families. In fact, the plants are so similar in appearance on the various steppes (as are the land forms) that if you were to transport even a relatively well-studied biologist to any of these regions, it is likely that they would not know which continent they were on for some time.

There does not appear to be a single published paper comparing the steppe climates of the four continents. Since the very birth and evolution of the human species has been so interwoven with all four steppe biomes from our very beginnings, it is all the more striking that so little effort has been made to try to quantify or compare the features of these regions.

Climatic Overview

Precipitation patterns and temperature isotherms are the two main factors that determine climate, and therefore the ecosystem that results from the interaction of climate and landform. Regions characterized as steppe receive, on average, 10 to 20 inches of rainfall annually; they experience protracted cold in the winter (temperatures dropping well below freezing nightly for three to six months) and likewise have hot summers, with temperatures often exceeding 30°C. Regions with hot summers that get less than 8 inches rainfall on average are generally considered arid (although some true deserts in warmer winter regions can receive precipitation comparable to steppe and are still regarded as desert). Geographers often use the terms "semi-arid continental climate" synonymously with "steppe climate," which by synecdoche becomes just "steppe."

The climate of steppe regions is a consequence of continentality: their central geographic position on their respective continents. All steppes are shielded from the prevailing westerly winds by mountain ranges or large expanses of land surface.

1. GRASSLAND STEPPE prevails along the eastern borders of all four regions: in North America this is represented by tallgrass prairie, which generally accrues the highest rainfall of any of the steppe regions (up to 30 inches in some years). Precipitation is heavily concentrated in the summer months. Grass is visually dominant in this form of steppe, perennial herbs being the second most visible element—especially plants in the Asteraceae during the summer and autumn months. Grassland contains a wide variety of other plant forms, including some bulbs, shrubs, and even annuals. But these are usually much fewer, both in terms of numbers of species and of individuals, than grasses and herbaceous perennials.

2. SHRUB STEPPE predominates in areas where there is greater precipitation (usually in the form of snow) in the winter months. In western America, this is typified by the Nevada sagebrush valleys. Here, not only sagebrush but many other shrubs—particularly in Rosaceae and Amaranthaceae—constitute the most obvious vegetative elements of the landscape. In areas that are not overgrazed, large numbers of bunchgrasses are usually present but also a much larger number and quantity of bulbous plants, annuals, and a diverse and sizable lot of forbs—perennials in many plant families.

3. MONTANE STEPPE occurs in high mountain valleys, parklands, and even above treeline in mountainous regions in all four steppe biomes. The islands of montane steppe often include xeric elements from the low-elevation floras nearest to the mountains, but these regions also have many high-altitude-adapted taxa more closely allied to more humid climates.

The Eurasian steppe lies to the north (in the rain shadow) of the highest mountain complex on earth—from the Caucasus in the west to the Himalaya in the east. The Altai mountains—which lie in the very heart of the Eurasian steppe—are equidistant from all four major oceans.

The Rocky Mountains lie to the west of the American summer-rainfall steppe, just as the Andes lie west of Patagonia. The highest mountain range skirts both hemispheres not far from the west coast, and the two hemispheres further mirror one another in many climatic features due to latitude. At higher latitudes the coastal rains result in temperate rainforests, while at lower latitudes they both transition through Mediterranean climates that become truly arid at the two tropics. The moisture borne by the westerlies comes mostly in winter and is dropped mostly on the western slopes of these mountains. Summer moisture in both North American steppe regions and the Patagonian equivalent comes mostly as usually brief summer convection storms cycling out of the Atlantic Ocean. The Patagonian steppe is much attenuated due to the constriction of South America toward the south, whereas steppe occupies a vastly larger and more complicated terrain in the American West and Midwest.

The South African steppe (consisting primarily of the Karoo and eastern highlands) is encircled to the south and separated from the Atlantic and Indian oceans by several ranges of mountains beginning in Namibia, extending south through the Richtersveld, Kamiesberg, Hantamsberg, and a plethora of ranges that transect the Cape Floristic Province—notably the Cederberg and Swartberg.

The higher escarpments of the Great Karoo extend eastward almost to the Drakensberg range, which barrier parallels South Africa's eastern coastline. The Drakensberg continues northward all the way to Swaziland and nearly to Zimbabwe; it is almost 1000 kilometers in extent and the highest mountain range in southern Africa.

Mountain ranges can have large impacts on their native flora. For one thing, mountains form barriers to the migration of plants. Any areas in the lee of the prevailing winds experience the rain shadow effect: rainfall is released on the west-facing slopes of mountains and then depleted by the time clouds crest the mountains, with much less precipitation occurring to the east. This phenomenon creates dramatic differences in precipitation extremes—sometimes over a very short distance—compared to regions where no mountains obstruct weather patterns. And mountains also generate weather: anyone who lives near high mountains can relate stories of uncanny and often dramatic weather phenomena, and on flat steppe or high mountains, such phenomena are often visible clearly and from afar. The highest peaks are often shrouded with their own weather systems—while the rest of the range can be basking in sun.

Chinook (or foehn) winds, among the highest sustained winds on earth, occur in the Rockies and Eurasia. When barometric pressure is severely depressed east of the ranges, the prevailing westerly winds accelerate and expand as they drop from the high pressure in the west to the east, sometimes approaching 200 mph for hours at a time. These winds can be quite warm in winter, sublimating snow and raising temperatures from freezing to 60°F or more. The desiccating effects of such winds in winter are considerable.

Hail can and does occur every year during the growing season. Many steppe climates have "hail alleys"—Cheyenne, Wyoming, for example, averages eight hailstorms a year. Most hailstorms are brief, with tiny stones that do relatively little damage, but many steppe regions have pockets prone to violent storms with substantial accumulation of hailstones. Apocalyptic storms with hail the size of softballs or grapefruits are not unheard of.

Untimely frost is a hallmark of steppe climates. Apricots, almonds, and other fruit trees are notoriously vulnerable to frost damage of buds or precocious flowers; in the Front Range, some trees bear fruit once a decade or less. The native flora has adapted somewhat to the fickle springs by emerging much later from dormancy than comparable floras in more equable climates.

Tornadoes are another characteristic weather feature of steppe regions: tornadoes can occur almost anywhere on the planet, including the oceans (sea spouts), but the weather patterns and topography of semi-arid grasslands spawn tornadoes on all continents except Antarctica. Grasslands obviously sustain far less damage from tornadoes than woodlands—and it's possible that tornadoes play a factor in the sparseness of woodlands in the "tornado belt" areas of the lower U.S. Midwest.

The vegetation of the steppes is a consequence of steppe climate, which in turn represents the dynamic interaction of semi-arid precipitation and extreme temperature gradients between summer and winter. The northernmost isotherms of steppe are correlated with the lowest precipitation rates (5 inches of rainfall constitutes steppe in Siberia), and conversely, the southern isotherm of steppe is where the greatest precipitation (up to 30 inches of rainfall) occurs. At northern latitudes, 30 inches rainfall would produce lush forests and at lower latitudes, 5 inches of rainfall would result in true desert. Indeed, the very existence of steppe is governed by precipitation (or lack of it), and precipitation on steppes can vary enormously, depending on the year and on the intensity of solar radiation. With such a wide spectrum, of both precipitation patterns and temperature isotherms, it's obvious that steppe can harbor a wide range of plants (especially when microclimate is factored in).

MICROCLIMATE

Microclimate impacts what plants grow where, in all climates. But the effects of microclimate in more arid and extreme climates are magnified dramatically. I recall a canyon in southeastern Colorado where acres of sunny prairie are filled with the sparsest of grasses and numerous xerophytic perennials, including many cacti, such as tree cholla (*Cylindropuntia imbricata*) growing nearly 8 feet tall. A local rancher invited us to visit her farm, where a deep gulley cut an escarpment nearly 50 feet high near the ranch: the north-facing cliff was festooned with maidenhair fern (*Adiantum capillus-veneris*), some with fronds 20 inches long, and the stream was filled with cardinal-flower (*Lobelia cardinalis*)—two spectacular mesophytes one would never associate with the southern Great Plains. Shady, moist microclimates throughout steppe climates often harbor relicts from more humid regions and times.

Conversely, canyons with protected, warm microclimates as far north as the foothills near Denver harbor xerophytic ferns that do not reappear until many hundreds of miles southward; for example, on Clear Creek canyon I found *Cheilanthes eatonii*, which does not commonly turn up until well into the Chihuahua Desert uplands in southern New Mexico. Such anomalies can occur in any vegetational system, but they seem to occur with great frequency in steppe due to the extreme gradient of temperature found on fully exposed slopes and shady aspects. Microclimate is magnified by the steppe dynamic.

MACROCLIMATE

The macroclimate, however, is what still informs steppe: the factors that are most important are the extreme contrast between summer and winter temperatures, the diurnal temperature extremes, and the paucity—and unpredictability—of precipitation. Anyone who has spent time over a period of years in steppe climates realizes the enormous differential in plant performance in dry years compared to wet ones. Researchers compared an "average" rainfall year to 1987 (a wet one): the same species bloomed in far larger numbers with longer season of bloom in the wet year. I recall that the extremely wet spring of 1983 resulted in *Calochortus nuttallii* showing up in tremendous numbers throughout Colorado's Grand Valley: hundreds appearing in median strips where they'd not been seen before. This sort of "flowering of the desert" phenomenon is characteristic of all steppe regions.

The most significant feature (aside from its paucity) about precipitation in steppe regions is that most steppes represent the transition between maritime climates to the east and more arid or Mediterranean climates to the south and west. Therefore, each steppe encompasses a subtle cline in timing of rainfall patterns. The precipitation is usually concentrated in the winter months at the far western end of each steppe (just east of the Sierra-Cascade crest in North America, just east of the Andes in Patagonia, and east of the Cederberg in South Africa). Eurasia is so vast that summer-rainfall steppes occur at the Volga River grasslands north of the Black Sea from Hungary extending eastward to the Caucasus, and then again from the Altai across Mongolia to China—with the winter-rainfall steppe, consisting primarily of shrubby vegetation and sparse grasses, sandwiched (as it were) between these extremes.

This geographic and climatic dichotomy—with winter rainfall at one end of the spectrum and summer rainfall at the other—leads to much of the confusion and misunderstanding surrounding steppe.

Geology of Steppe

Just as the steppe climate is profoundly influenced by the proximity of mountains, the very processes that created those mountains have shaped the chemistry and substance of steppe soils—the matrix upon which all plants depend. Although the age of mountain-building episodes in each of the four steppe regions is radically different and largely unrelated, there is a fundamental parallel in the way in which ancient seabeds and the upthrust of mountains interacted to shape present-day steppe landscapes and produce the complex bedrock formations.

Geologists have postulated that present-day mountain ranges arose where ancient erosion deposited enormous depths of sediments, compounded by subduction (the process where crust is folded deep into the mantle along the seams where continental plates collide). Once sufficient depths of lighter, crustal rock are accumulated, deep currents in the mantle can force the rock of lesser specific gravity upward, forming a mountain range in time. Something along these lines is probably responsible for orogeny in western America, the Andes, and South Africa. The dramatic and relatively recent orogeny of the Himalaya, Alps, and Caucasus are undoubtedly due to the impact of colliding plates—India colliding into Asia resulting in the Himalaya, and North Africa and Arabia colliding with Europe to produce the Alps and Caucasus.

In Europe, the Himalaya, and the Americas, large layers of sedimentary rock are still present on and around the highest peaks. The plains around the mountains and even at the highest elevations consist of these deep layers of limestone, sandstone, and shale, deposited by Mesozoic seas. In southern Africa the maritime deposits are much smaller and largely on the peripheries of the continent. Rock and soil is flamboyantly displayed in the sister steppe climates of Eurasia, Patagonia, and especially the intermountain American West; in the Colorado Plateau, a dozen or more national parks and monuments and countless acres of BLM and other public lands (and a good deal of private land as well) display a hypnotic array of natural sculpture that is a major tourist attraction.

The core of these mountain regions is generally igneous, varying from andesite in the Andes, to variations on basalt, granite, and quartzite in the rest. Many regions are intermediate between sedimentary and igneous—with metamorphosed limestone forming marble and sandstones and shales forming gneiss, schist, and quartzite in greater or lesser proportions. The chemistry and mechanical properties of these rocks as they near the surface combine with the action of climate to form the remarkably diverse soil types that characterize steppe regions.

STEPPE SOILS

A wetter climate's minerals are often leached quickly from soils, but steppe soils are often very rich in surface minerals, and soils can vary radically in texture and profile over a very short distance. Selenium is often concentrated in semi-arid regions, making nearby groundwater undrinkable. You can often smell its pungent odor on soils where it is concentrated, and a spectrum of selenium-tolerant plants will replace more widespread but less tolerant flora in the region.

Various shales and sandstones weather to produce extremely complex and chemically rich clay soils that possess unique qualities—especially when compared to soils derived from clays in wetter regions. The profound effects of sun and deep winter frosts cause steppe soils to flocculate—a physical process that clumps and organizes clay plates into geometric arrays that are maintained by electromagnetic and chemical reaction. Flocculation results in quantities of air permeating deeply into the soil profile: the complex chemical reactions caused by air, water, and soil in semi-arid regions yield soils that are enormously permeable by water compared to the compacted clays of maritime and subtropical regions.

The clays in semi-arid regions are characterized as montmorillonite clays (or closely allied bentonite clays), which can absorb enormous quantities of water. When their tilth (the ideal, fluffy texture of soil that has not been trampled) is maintained, they disperse this water quickly and efficiently to great depths. Their surface can dry quickly in the sun, but moisture is preserved at depths, for long periods, even during droughts.

Unlike kaolinite clays in humid climates, montmorillonite clays maintain a very high cation ratio that binds nutrients essential for biota. The seemingly sterile steppes, as a consequence, contain a very rich soil web that in turn (if not destroyed by compaction) sustains a similarly rich fauna and flora above ground. Soil scientists have characterized many steppe soils as pedocals (as opposed to the pedalfers of more humid regions). Pedocals often contain much higher proportions of calcium from the alkaline rock substrates (especially limestone and calcareous shales), which can cause the soil pH to rise to 8 or more. Similarly, calcareous rock in more humid climates is often leached of its alkaline chemicals by greater rainfall, resulting in much more circumneutral soils.

Humanity has tried to replicate the chemistry of steppe soils in wetter regions. Gardeners in heavy-rainfall areas of New England or the upper Midwest are often told to add calcium carbonate or crushed dolomite (magnesium carbonate) to their soils in order to "sweeten" them and encourage the growth of grass on lawns, or productivity in vegetable gardens. The need to raise soil pH in humid regions is a sure indicator of the steppic origins of much of our garden flora; many classic groups of vegetables, in particular, perform poorly in the intensely acid soils of wet climates.

Soils are a result of the interaction of climate and substrate interacting with micro- and macroorganisms. Since the surface vegetation in steppe is often relatively sparse, soils and often their rocky substrates are clearly exposed and surprisingly variable over a short distance. Steppe shares with desert climates the exposure and immediacy of geology: in humid climates the rocky source of soil is usually obscured beneath vegetation—tempered by precipitation and lush growth.

MOUNTAIN-BUILDING AND FLORISTICS

All steppe regions abut onto a Mediterranean biome along their westernmost flanks: since this is so on all four major continents, this is clearly due to the prevailing westerly direction of storm movement and the mid-latitudes where both climate regimes occur (at their very widest, between the tropics and the 50th parallel in both hemispheres). The gradient between Mediterranean biomes and their "steppe sisters" would appear to be a function of the greater continentality of steppe climates. This continuum can be observed in South Africa where the Renosterveld ("rhinoceros fields," clay shrublands where rhinoceros once roamed) found amid the more truly Mediterranean Fynbos are interconnected between mountains, gradually expanding into the Worcester Karoo and ultimately the Great Karoo beyond. As a consequence, there is a great overlap of plant families between Mediterranean and steppe regions. In fact, there are highly polymorphic species that seem to grow indiscriminately in both Mediterranean and their neighboring steppe biomes.

The great overlap of each steppe flora to the neighboring Mediterranean flora (despite the seeming opposition of their respective climate regimes) seems to suggest that there was a continuum extending from coastal Mediterranean to the most continental steppe at some time in the past, and that plants may well have adapted and evolved in situ subsequent to the rise of the Rockies, Andes, Caucasus, Himalaya, and Alps, separating closely related taxa in space. In particular, the disjunction between coastal Mediterranean plants and their alpine correlatives in the Himalaya is so frequent in different families and genera that it seems to confirm the existence of a continuous range prior to orogeny.

Altitude and Steppe

Although in the popular imagination, steppe suggests flat grasslands, the various elements that result in steppe vegetation (climatic extremes, reduced rainfall, pedocal soils) occur in a wide variety of geographic terrain and altitudes. I once had the privilege of leading a dozen Russian botanic garden directors down the Mount Goliath trail in Denver Botanic Gardens' alpine unit. It was late June, and the flowers were at their very peak of glory. At one point, I referred to the surrounding "alpine tundra," and one of the directors promptly corrected me: "This is not tundra—this is alpine steppe!" I recall to this day my shock and amusement as I watched all the others nod vigorously in agreement. Since both "tundra" and "steppe" are Russian words (as is "taiga," for that matter) we would do well to recognize the experience Russians have with the primary biome of their vast country: Russians know steppe!

Tundra is characterized by permafrost (and does occur higher up, on Mount Evans). The south-facing slopes of Mount Goliath are much warmer; they are covered with a fairly dense mat of various grasses and sedges and a surprising number of forbs, closely related or identical to taxa growing on the foothills and the plains; for example, you can find *Penstemon virens* and *Heuchera bracteata* (ordinarily restricted to the lower foothills, below 8000 feet) on rocky outcrops at nearly 12,000 feet on the mountain here.

There are places in Patagonia and Texas where steppe-like vegetation reaches down to near the seashore, and it ranges not only over vast areas like the Great Plains but onto mountain meadows and parklands at lofty elevations, all the way to tundra—wherever sparse rainfall is accompanied by severe cold in winter and heat in the summer.

Origins of Steppe Flora

The largest families of plants that now dominate steppe floras around the world—Asteraceae, Poaceae, Brassicaceae, and Scrophulariaceae—are believed to have arisen and evolved fairly late in the current Cenozoic era, long after much of the mountain-building in steppe regions was under way. Since herbaceous plants—especially those from xeric or alpine environments—are rarely preserved as fossils, much of the paleobotanical data is inferred by genetics and current distributions. It is likely that distant ancestors of these large families may well have occurred in more than one of the steppe regions: a significant proportion of the floras of North America and Eurasia are so closely related that they are believed to derive from a common floral ancestry from times when these continents were conjoined—or at least connected. The twinning of *Abies* and *Picea* at subalpine latitudes across the northern hemisphere suggests their origins may be traced to a common ancestor that once grew in the highlands of Laurasia. Myriad other comparable twinnings between continents argues for the uniform genetic origins of much of the highest latitude north temperate flora (termed Holarctic by botanists and zoologists alike), their distribution often characterized as circumboreal.

A significant proportion of northern hemisphere steppe flora consists of classic montane and alpine families that adapted to drier or lower environments. In Gentianaceae, for instance, *Gentiana olivieri* is a drought-adapted gentian of Central Asia, just as *G. bigelovii* and *G. puberulenta* have adapted to steppe habitats in North America; and in western America, *Frasera albomarginata* and *F. coloradensis* have adapted to grow on dry steppe—undoubtedly they evolved into xerophytes in situ as their habitat grew ever drier since the Miocene. The uncanny parallels between so many plant families argues that some sort of commonality must have existed at some point.

Parallels in Predominant Families of Steppe Flora

ASTERACEAE

The 12 tribes of Asteraceae recognized for most of the past two centuries have morphed into 12 very different subfamilies, five of which are relatively large. The cladists who have analyzed the molecular structure of the daisy family have determined that three of these five subfamilies represent the most recently evolved and most widespread members of the aster family on all continents.

The South American subfamily Mutisioideae has several steppe representatives: widespread throughout the Andes is the namesake genus *Mutisia*, many species of which form large vines with very showy flowers; the other genera are not as important horticulturally. The thistle subfamily (Carduoideae) is notorious worldwide for its many pernicious agricultural weeds. The intensely spiny nature of this group is characteristic, as are the shaving-brush flowers. Of course, most thistles are not particularly weedy, and some are extremely ornamental. The greatest concentration of species in Carduoideae are in Eurasia, with a second, smaller concentration in North America.

POACEAE

Grasses visually dominate most steppe regions, although the species numbers of grasses may not always equal the number of species found in other groups in various steppe climates. The northern hemisphere has much more extensive prairies than the southern, and there is a considerable similarity in the species types and habitats between New and Old World steppes. *Stipa* is especially widespread in Asia, although it can be conspicuous and abundant in the American prairie as well. Many genera are abundant in all four steppe regions, for example: *Sporobolus albicans* (South Africa), *S. rigens* (Patagonia), *S. airoides* (North America), and *S. balansae* (Eurasia).

BRASSICACEAE

The diversity of taxa in the mustard family is greatest in Eurasia but still significant in North America; both regions share a number of important genera in this family, but there are many more endemics in Eurasia than in North American steppe. The family has a much smaller but still visible presence in the southern hemisphere.

SCROPHULARIACEAE

In the American steppe, *Penstemon* and *Castilleja* are the two largest genera in this family; both are present in Eurasia as only a single species. *Veronica* has significant representation in both hemispheres. South America's chief representative is *Ourisia*, which contains many very decorative alpine and montane representatives. *Verbascum* and many other smaller genera are associated with the Mediterranean. *Diascia* and several others of the numerous southern hemisphere endemic genera bear a surprising superficial resemblance to northern hemisphere scrophs, although they undoubtedly have ancient provenance in the south. Many former members of this family are now placed in Plantaginaceae and Orobanchaceae.

FABACEAE

The third largest family of plants is diverse and numerous in all steppe regions (and equally common in the Mediterranean biomes and throughout the tropics). Members of the pea family make up a significant amount of the local species in all steppe climates. They are also a prominent visual element in the landscape. *Astragalus*, *Oxytropis*, and *Lupinus* are significant in both New and Old World steppes.

ROSACEAE

The Rosaceae are highly diversified in both North America and Eurasia, beginning with *Rosa* but also including *Malus*, *Prunus*, and other woody genera that are widespread in both hemispheres. An assortment of strange, xeric Rosaceae are especially abundant in North America (*Cercocarpus*, *Fallugia*, *Chamaebatiaria*); some appear to have what must be ancient relationships with *Spiraeanthus*. *Agrimonia* is found in North America, Eurasia, and South Africa; *Alchemilla* occurs in South Africa and is very common across all Eurasia and South America (but strangely absent from North America north of Mexico).

POLYGONACEAE

Prominent in northern hemisphere steppe: *Eriogonum* is one of the largest endemic genera of North America, distributed from coast to coast but most common west of the Mississippi, where it is sometimes a dominant element in desert, Mediterranean climate, and steppe landscapes from the coast to the highest peaks. In Eurasia, *Polygonum* and *Persicaria* are widely dispersed and prominent, as are *Rumex*, *Rheum*, and *Atraphaxis*, which shrubby genus superficially resembles many of the more fruticose eriogonums.

AMARANTHACEAE

Formerly Chenopodiaceae. *Atriplex* is extremely diverse and prevalent in highly alkaline substrates in all hemispheres. *Krascheninnikovia* is the classic twinned genus, linking western America and Central Asia.

APIACEAE

There are dozens of endemic genera in both hemispheres, and some remarkable twinned species that probably derive from the same Holarctic ancestor.

RANUNCULACEAE

Ranunculus and *Anemone* occur in all four continents, and some taxa suggest a very ancient common ancestry between continents now separated distantly.

BORAGINACEAE

As with Brassicaceae, especially richly diversified in Eurasia. *Myosotis* is present in all four steppe regions, and rich speciation occurs in *Cryptantha* in western America.

APOCYNACEAE

Now includes Asclepiadaceae. Genera are widely distributed in all four steppe regions, although the genera in each region are often distinct.

MONTIACEAE

Formerly included in Portulacaceae, this family occurs in all four regions—*Claytonia* is represented in much of the northern hemisphere, although closely allied to *Calandrinia* (which occurs in both North and South America).

MALVACEAE

Recent research has combined Sterculiaceae with this family. Malvaceae, although not a large family, is nonetheless diversified and quite numerous in all four steppe regions, for example, *Sphaeralcea* in western America and South America, *Alcea* and *Althaea* in Eurasia, *Hermannia* in South Africa (and Texas), and *Cristaria* in South America.

LAMIACEAE

The mint family is very well represented in all four regions. *Salvia* occurs in all four hemispheres, and many genera, including *Scutellaria*, occur in three of the four steppes. *Satureja darwinii* from South America is matched with similar taxa throughout Eurasia.

FRANKENIACEAE

Frankeniaceae consists of compact, mat-forming groundcovers and dwarf shrubs widely distributed in all four steppe regions. Most species grow in sandy habitats, or on highly alkaline outcrops. Many species are found on the saline flats around lakes or the seashore. *Frankenia patagonica* is restricted, as its epithet suggests, to South America. *Frankenia pulverulenta* is widespread in the Karoo of South Africa. *Frankenia jamesii* is a shrubby xerophyte largely restricted to rocky outcrops in the Arkansas River Valley on the Great Plains of Colorado, extending southward spottily to Texas. *Frankenia bucharica*, found in Central Asia, looks uncannily like the various southern hemisphere and Mediterranean species.

POLYGALACEAE

Polygalaceae has rich representation in all four steppe regions (and their neighboring Mediterranean biomes as well). Although a great deal of variation in habit occurs within the family, some uncanny similarities exist between widely separated taxa: on the Roggeveld escarpment of South Africa, the tufted, pink-flowered *Polygala scabra* has developed a spiny habit surprisingly similar to *P. spinescens* from Patagonia and *P. subspinosa* from the Colorado Plateau. The genus is widely distributed in Central Asia as well, where we have seen *P. comosa* at many elevations in the Altai of Kazakhstan and Mongolia. These sorts of striking parallels are found in many more families.

Plant-People Connection

Early primates evolved at the fringes of the grasslands of southern Africa; they were largely vegetarian, as our cousins, the chimpanzees, still are. Hominids evolved into the genus *Homo* only between two and three million years ago, when they ventured permanently as carnivorous hunters onto grassland—where now only rodents, ungulates, and pachyderms survive as vegetarians on the fibrous, cellulosic grasses. It can be said that our generic leap from ape to human was a consequence of our change of habitat from forest to grassland. Ever since, steppes and savannas of the world continue to nurture and alter our species. This vegetation has served as the larder, medicine chest, and superhighway of human migration since our very beginnings. The flora of the steppe has provided healing herbs, roots, and greens for human consumption for countless millennia. That flora was responsible for the vast herds of ungulates that were the primary prey and protein source of our ancestors. Humans have crossed continental boundaries and sailed vast seas in pursuit of prey and fresh habitat, ultimately becoming the pantropical, universal phenomenon we are today.

Humans occupied steppe and their tropical equivalent savanna climates almost exclusively for the millions of years we were evolving into our current form. Although our population grew when we forsook the forests for grassland, it was not until crops were domesticated, around 10,000 BC, that populations began to explode. Once again, this revolution was a consequence of the steppe biome and its flora. A handful of species of steppe plants gave rise to those first cultivated crops. True, humans have subsequently adapted wheat, corn, and other grains to grow in more humid climates. For steppe crops to succeed, however, the forests that prevail in humid climates had to be felled in order to create a steppe-like environment—the vegetable garden or cereal plot. To this day, farming and gardening in wet climates necessitates heavy fertilization with minerals to approximate the balance that occurs naturally in steppe soils—hence the frequent "liming" of acid soils, so that their pH approximates the alkalinity of steppe pedocals, which are naturally high in calcium carbonate. The great explosion of humanity's population subsequently took place in more maritime regions, with their more predictable rainfall and more temperate climates, and the awareness of our steppe beginnings has largely been obscured, or simply forgotten. The vast steppes (almost the only habitat occupied by humans for millennia) largely became backwaters of humanity—underdeveloped, sparse in population, often overgrazed, but otherwise much the same as they had been throughout prehistory.

Prior to the "rediscovery" of steppe in the 20th century, all four steppe biomes were largely the province of pastoralists: in North America the steppe had been occupied for 10,000 years and more by Amerindian tribes who depended on the vast herds of ungulates for sustenance. These herds were displaced by cattle in the 19th century, and the cowboy became, perhaps, America's most iconic hero. This same process occurred in South America, with the gaucho becoming a similarly iconic representation of Patagonia and the pampas. Pastoralists are still the dominant residents of much of the Eurasian steppe: let's not forget that Genghis Khan and Tamerlane's soldiers were herdsmen when not on the warpath. Prior to European colonization, the only native peoples in much of the southern parts of Africa were a distinctive race (often referred to as Khoi or San) with cinnamon-colored skin and delicate build and features. In the deserts they constituted bushmen—but in much of the slightly moister grasslands, the Khoi were and are pastoralists with herds of cattle.

As humans maximized and perhaps exceeded the carrying capacity of humid and tropical maritime climates, it's not surprising that they have rediscovered the promise of the harsh steppe environment. Cowboys are not very often encountered in cities like Denver, Omaha, or Salt Lake City, which have burgeoned in recent decades—the so-called Sun Belt population explosion. A comparable boom is occurring across the former Soviet Union's steppe nations: from the Caucasus throughout the "stans" to Mongolia, cities are expanding and populations are rising again where Genghis Khan and Tamerlane marched their fierce armies.

In the 21st century, the various steppe climates have become a geopolitical focus for energy development of all kinds: from fossil fuels to solar and wind power (which can be maximized in these sunny, windy high regions). And many countries in western and Central Asia—Iraq, Iran, Afghanistan, Pakistan—are literal hotspots, politically. Argentina has returned to a somewhat more stable status after the political turmoil of the Peron years and a decade of economic collapse, although Eva's ghost can be heard rattling a sabre whenever the Falklands are mentioned. More and more economists predict enormous capital investment and growth in Africa, the poorest continent, in this century: this transformation is likely to be spearheaded by South Africa, by far the wealthiest, most dynamic economy on that continent. But just as protohumans may have taken to bipedalism on the Highveld, their descendants are grappling with atavistic tribal animosity and a powder keg of struggle over the newly discovered resources of the steppe. The new steppe warriors are the frackers and drillers, perhaps—and the land developers.

Our psychology, ecology, and habits as *Homo sapiens* were undoubtedly shaped by the millennia of evolution that took place exclusively in steppe and savanna environments, where our existence depended on our efficiency and ruthlessness as hunters (and our intelligence evolved rapidly to help us avoid becoming prey). Perhaps our rediscovery of steppe will allow humans to utilize this same intelligence to silence the primal tendencies to violence in our nature, and to design a future relationship with natural environments that is mutually sustainable.

Humanity is experiencing a homecoming of sorts, as our future hinges more and more on the political and cultural dramas taking place on the world's steppes. In the process, the fragile landscape, once graced by vast herds of ungulates and sparse tribes of humanity, is transforming rapidly and radically thanks to resource extraction, marginal farming, and geopolitical wrangling. The warrior has returned to the steppes armed with a panoply of technology this time. Are we fated to continue our role as rapacious exploiters, or shall the steppes help us become wise stewards this time around?

THE CENTRAL ASIAN STEPPE

MICHAEL BONE

CENTRAL ASIAN STEPPE

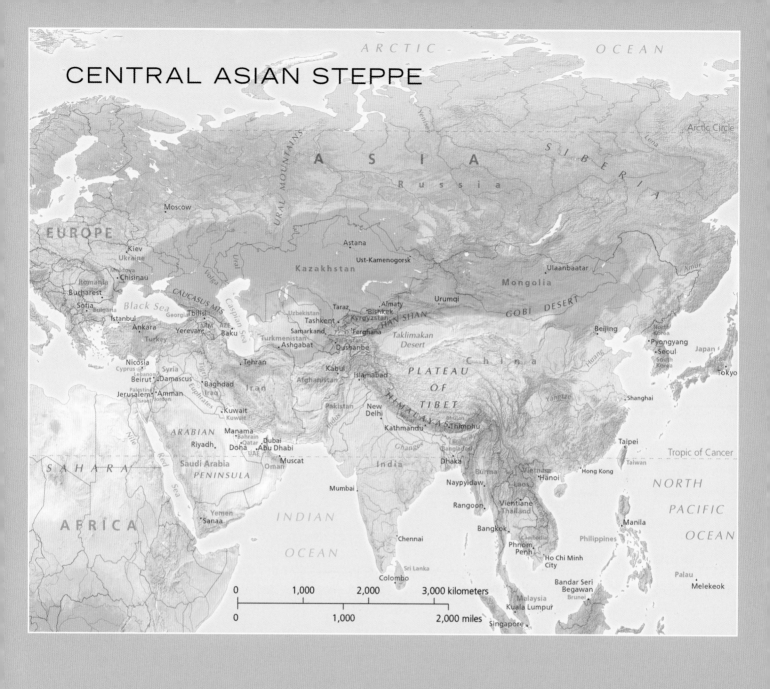

THE GREAT STEPPE of Central Asia has been the stage for much of the drama that is the human experience. The foods we eat are products of the steppe. Travel and human migration, as well as the domestication of animals, enslavement of races, and vicious fights for freedom and survival—all are products of the steppe. Kingdoms have risen and fallen here; religion and spirituality often and still dominate how humans control its pathways. This is a land of discovery, where we humans have had a constant presence and yet have forgotten our relatively recent past. This expansive part of the world helps to define us, yet we have always struggled to understand that at the core of our evolution as a species, we are children of the steppe. Perhaps the ancestral lands we trod for thousands of years feel trite and understood, while the distant forest or jungle seems to be the place that looms with mystery. But the steppe holds its own set of mysteries. Even the concepts of physical boundaries in this region of the world are greatly debatable. Changing weather patterns and floristic diversity (Kazakhstan boasts 6000 taxa) blur its edges. Throughout the steppe and the various mountain ranges that define its borders, the greatest number of species—and the most interesting ones—are most often found where mountains meet the steppe. Numerous pressures are at play, together creating this vast and beautiful region, the steppe of Eurasia.

Both the North American and Eurasian steppes fall in the Holarctic Kingdom, according to the division of Armen Takhtajan (1986). The Central Asian steppe contains the Boreal and Tethyan subkingdoms; Takhtajan further divides the Eurasian steppe into two floristic regions: the Circumboreal and the Irano-Turanian, each aligned to the preceding subkingdoms. The Circumboreal is the largest region outlined by Takhtajan; he goes into it in great depth, further subdividing this area into 15 unique floristic provinces. In many respects, America and

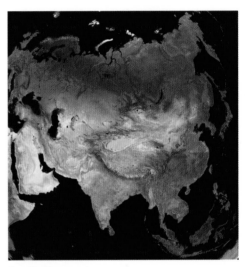
Satellite image of
Central Asia.

OVERLEAF Starry skies over
Ikh Nart Nature Reserve.

Paraquilegia caespitosa, the evolutionary precursor to the columbine.

Eurasia share a great deal of the same floristic pressures dictating the evolution of the plants that surround us. It is easy to taste the similarities and see the distribution patterns of many of these plants. The circumboreal floristic provinces that influence the Eurasian steppe in the greatest way are the eastern European, western Siberian, Altai-Sayan, Transbaikalian, and the northeastern Siberian provinces. The other provinces tie the look and feel of these areas together by sharing many of the same important families and genera, but because of their more stable temperature and higher annual moisture events, only pockets of steppe are expressed. The northernmost provinces consist of taiga forests and frozen tundra, through Siberia and across the Bering Strait to the upper Canadian provinces. All this gives us that feeling of a contiguous flow of plants, and the visual similarities help to lump many of these groups together.

Where the Boreal and Tethyan subkingdoms collide is where the allure of the exotic and the tug at our long-forgotten forager/nomadic heritage can be found. There in that transition, the lines of familiar are blurred and a hybridizing of the present with an ancient past occurs. Indeed, one can almost follow the exact line drawn on Takhtajan's map, and see the paths both of the Central Asian steppe and that of the Silk Road. In the Tethyan Subkingdom, we lose the familiarity of what we are used to seeing in North America and find the mystery of exciting and different genera (expressions like *Acantholimon* and the endemic *Paraquilegia*) and even families. As the two subkingdoms merge, the familiar blends with the exotic and strange.

Geography and Climate

The Eurasian steppe is the largest habitat of the Eurasian supercontinent, falling approximately between 40° and 55°N, beginning roughly at 28°E in Moldova and ending at 125°E, where the Gobi Desert meets the Siberian tundra. That makes the total area of the contiguous Central Asian steppe approximately 685,000,000 hectares. With modern tools like satellite imaging, we are able to look at the interrelationships of habitats literally from the palms of our hands. The theories of plate tectonics and continental drift play crucial roles in defining the boundaries of the steppe. In Central Asia, one such defining "moment" was when the Indian plate smashed into southern Asia, forming the Tien Shan and the Himalayan complex of mountains. Another such concerns the Arabian plate, which presses into southern Asia, creating dozens of smaller mountain ranges that run along the east-west border and separate the oceanic weather influences, thereby creating rain shadows, which in turn create the steppe. The Mediterranean Sea to the west sees the limit of its climatological influences moving east as you approach the Caspian Sea. Major weather patterns are also influenced by the Indian Ocean, but most of the rainfall hits mainland India and is stopped by the Himalaya. North of the Himalaya lies the great Tibetan Plateau. From the north, moisture is raised from the Arctic Ocean and deposited over Siberia and the taiga forests of Russia, leaving little rain by the time weather systems move south and reach Kazakhstan. These factors pushing from both north and south have created a long corridor of steppe reaching from the Caspian Sea to the Pacific Ocean.

The Eurasian steppe extends from west to east, starting in Moldova and then passing through the heart of Ukraine and continuing on to surround the Caspian Sea. In between the Black and Caspian seas, the Eurasian steppe meets the Caucasus. From there, it stretches all along South Kazakhstan and continues to the Altai and through Mongolia, all the way to its western edge. The steppe is bordered to the north by Siberia and the taiga forest. Even in the northernmost boundaries of what we traditionally call Siberian tundra, we find the low grassland steppe scattered throughout this cold northern plain. In Mongolia, the steppe is bordered on the south by the vast Gobi Desert. South of there, pockets of steppe can be found even along the heights of the Tibetan Plateau. In fact, the steppe is so ingrained into every aspect of life in Central Asia that they have names for at least seven variations on the theme: true steppe, desert steppe, montane steppe, forest steppe, alpine steppe, shrub steppe, and meadow steppe.

LAKE BALKHASH

The major water reserve in central and eastern Kazakhstan is Lake Balkhash, which sits at the low spot of an endorheic basin. This water body is unusual in that it is separated into very different halves: the western half, which is fed mainly by the

Montane steppe on the Mongolia-China border.

Ili River, is freshwater; the eastern, fed largely by the Ayagoz River, is very saline. The evaporative water movement from this lake is sent mostly to the corner of the Tien Shan and is deposited in the mountains east of Almaty and along the Chinese and Kyrgyz borders. Lake Balkhash is very important as a major stopping place for migratory birds (on their way from Siberia to India and vice versa), but as with the Aral Sea, increase in population and higher demands by irrigated croplands have made the situation for this large body of water very unstable.

LAKE ISSYK KUL

Issyk Kul is another important endorheic basin lake that moderates and influences the climate in the southeastern portions of the steppe: crucially, it is located on the southernmost expression of steppe, south of the Tien Shan and Himalayan complexes. The evaporative pressure of Issyk Kul is the force that dominates weather patterns. This area is extremely diverse, thanks to a unique set of evolutionary floristic pressures that create speciation. In portions of Pakistan are some very interesting expressions of high montane steppe and some equally interesting plants. *Acantholimon* has a high level of speciation in this area, forming tight hummocks up to 6 feet across and 18 inches or more tall at near alpine elevations.

Alpine steppe in Pakistan.

Meadow steppe in Kazakhstan.

Forest steppe in Mongolia north of Ulaanbaatar.

The Karatow.

Visually these resemble the *Junellia* or *Bolax* species of Patagonia. High winds and extreme cold help shape the forms of these plants.

THE KARATOW

The Karatow are believed to be one of the oldest mountain ranges in the world. There are no high peaks or jagged cliffs, but there are unique habitats and some ancient populations of plants. Here in the decomposing mountains are still wild herds of Bactrian camels, which are most commonly associated with the Gobi Desert. Here, where the directionally north-south Karatow meet the directionally east-west Tien Shan, some of the interesting floral components are bulbs, *Iris*, and *Rheum*. *Rheum tataricum* truly stands out as unusual in this area, with large flat leaves hugging the earth, trapping moisture that would otherwise evaporate. Flower stalks of *R. tataricum* can stand as tall as 2 feet, and once dried, they break free of the dormant plant and tumble across the steppe, collecting in small ravines and washes by the thousands. On rocky slopes, which afford enough disturbance to limit competition, you can find *Iris willmottiana* or *I. zenaidae* in early spring bloom. Several species tulips occur in the Karatow, including *Tulipa bifloriformis* and the king of the steppes (and all tulips), *T. greigii*.

There are very few roads through here; except for the main highway, most all roads are dirt or just tracks in the dry expansive steppe. On occasion you can find a spring or a low salty lake where there is a little more moisture to encourage a broader speciation, and you can see the dominant shrub of the lowland steppe, *Atraphaxis karataviensis*. The farther north you move from the Tien Shan, the

Rheum tataricum.

more the species of plants in the Karatow change; there is a very high degree of endemism with special plants like *Aquilegia karatavica*, *Delphinium pavlovii*, and as many as 16 endemic species of *Astragalus*.

On the Karatow's southern tip, the Tien Shan are exerting a much newer evolutionary pressure on the flora, which has remained relatively stable from the time of the Tethys Sea. One example of this is the rare *Spiraeanthus schrenkianus*, an obscure member of the rose family that shares a striking visual similarity with the North American Great Basin native *Chamaebatiaria millefolium* (which happens to not be rare): on a visit to Denver Botanic Gardens, a dear friend and talented naturalist from Kazakhstan marveled at the chamaebatiaria and demanded to know how in the world we were able to get seed and grow the plant he knows as spiraeanthus. To this day, he is convinced the two plants are one and the same—only populations disjunct by some 9000 miles. Toward the north end of the Karatow the flora is much more influenced by the Qizilqum desert to the west and the Betpaqdala desert to the east. Any moisture that develops in the Aral gets deposited on the Karatow, giving a chance at life in an otherwise very harsh habitat.

TIEN SHAN PROPER

If you were to travel the great Silk Road from southern Europe to China, you would visit ancient cities like Ashgabat, Bukhara, Navoi, Jizzakh, Shymkent, Bishkek, and Taraz—all with stunning views of the great Tien Shan mountains, all situated to take advantage of fresh clean water from those same mountains. Of

An early 12th-century burial tomb, just east of Taraz, is one of the few structures not destroyed by the Mongol hordes.

RIGHT Limestone hoodoos in Tien Shan proper, much like those in the Badlands of South Dakota.

all those cities, I find Taraz, recently recognized as the oldest city in Kazakhstan, to be the most interesting. The town, strategically set at the base of the Tien Shan and Karatow mountains, along the Talas River, has been a trading center along the Silk Road for thousands of years. Every so often a new culture would rise and try to take control of this city, or just burn it to the ground; it remains, however, a great location from which to explore the Tien Shan mountains.

The flora along the northern border of the Tien Shan is very different from the southern side. Along the northern border you see much more of the expressions of the Boreal Kingdom. Shortgrasses dominate visually, and as geology changes, so do the subtle expressions of plant life. Along each of the rivers as you approach the foothills, you find wonderful pockets of plant life. In a low meadow near the border with Kazakhstan and Kyrgyzstan, *Eremurus fuscus* and *E. robustus* create a natural hybrid population where every plant is unique. Size and shape is wonderfully intertwined with the native prairie. At the low point in this field is one of Central Asia's unique floral tie-ins with South African flora, the elusive *Ungernia sewerzowii*, from the amaryllis family. Tall spikes with multi-headed flowers in browns and burgundies make this plant truly stand out among the grasses. Not far from the site of the ungernia is a valley that is literally the border with Kyrgyzstan and Kazakhstan. The valley is made up of wonderful composite limestone hoodoos. Growing among them is a low-elevation population of *Saxifraga albertii*. Normally this species is found at much higher and colder locations; here, however, the plants cling to the vertical faces of this canyon at just the right elevation to capture the seasonal water that storms through but not so low as to be tumbled down and away by rushing currents. The base and floor of this canyon is carpeted with *Atraphaxis* species, in hundreds of variations of height and color, from bright candy-apple red to pure white. Many of these shrubs carry a honey-sweet fragrance; others smell of sour milk or wet woolen socks that have been worn too long inside a flooded wellie.

THE ALTAI

The Altai is a conglomeration of 53 smaller mountain ranges, most running on a north-south axis. Climate around the Altai is varied but can be generalized as follows: to the east is shortgrass steppe with cold, dry winters; to the north is the world's largest taiga forest, with ample winter snowfall; to the west there is again steppe, only with slightly more annual precipitation and mixed-grass prairies; to the south is the Gobi Desert. Ust-Kamenogorsk, nestled on the banks of the Black Irtysh River, on the flat steppes below the Altai, is another prime location from which to experience the Central Asian steppe. Here, just out of the city proper, rich rolling hills support an intense floristic area. Hillsides are covered with the showiest of flowers—*Clematis integrifolia*, *Nepeta sibirica*, *Paeonia tenuifolia*, the endemic *Daphne altaica*—and in the shortgrass steppe are rocky outcrops, where you can find plants like *Orostachys spinosa* and *Lilium martagon*. There are also roses in this area, making what looks like a formal garden with shrubs, perennials, and well-placed rocks. On the ridgetops you see plants like *Rheum compactum*, *Eremurus fuscus*, and *E. altaicus*.

As the rolling foothills continue you find some interesting trees—*Acer tataricum*, an interesting population of Scots pine (*Pinus sylvestris*)—but little to no great expressions of forest. It is not until you get deeper into the heart of the Altai that any type of forest occurs. The diversity is not very great under the forest canopy, but nearby clearings offer a host of fun vegetation, and near "mosquito lake," a population of *Allium tulipifolium* punctuates the feather grass and *Phlomis oreophila*.

The Black Irtysh River has been dammed multiple times to provide irrigation and water for the cities, but some areas along it are still pristine—for example, there are wonderful sand hills with beautiful cushions of *Astragalus* as well as *Clematis songarica*. Leaving Ust-Kamenogorsk you must cross the Irtysh River to get into the Altai proper. Once past the river you begin to really get the feeling that you are in the mountains. There are areas of spruce and birch forest and wonderful alpine meadows. Beyond some of the smaller ranges are intermountain depressions and pristine intermontane steppes.

In the Mongolian Altai there again is a slight shift in the flora of the intermontane steppe. On this side you begin to see *Caragana* species as the dominant shrub vegetation, mainly because they are spiny and not palatable to camels

Ungernia sewerzowii in true steppe along the Silk Road.

Clematis integrifolia.

or goats. Among the more noteworthy peashrubs are *C. altaica*, *C. arborescens*, *C. bungei*, *C. microphylla*, and *C. spinosa*. Of these, *C. bungei* and *C. microphylla* are easily the dominant species. This eastern portion of the Altai tends to be greatly overgrazed. Many of the herdsman drive their goats into the mountains looking for lush valleys and water sources; bound by political borders, they no longer have as much acreage to find fertile grazing grounds. Pristine areas are tucked deep into the mountains; one must cross high and treacherous passes—places even goats won't travel—to get to rich alpine meadows, irrigated by melting glaciers and teeming with wildflowers.

MONGOLIAN STEPPE

The most common conjuration of the term "steppe" is that of the vast Mongolian steppe, birthplace of Genghis Khan and the legendary Mongol hordes. The steppe here is predominantly true steppe and desert steppe. A large portion of Mongolia is the Gobi Desert, which dictates the great steppe north of it. The major floristic pressure is of the Boreal Kingdom. To the far east and to the south the pressures of the Himalayan flora create endemism. The landscape is very interesting, with

Overgrazed steppe in the Mongolian Altai.

LEFT Markokol Lake, high in the Altai mountains, is home to a rare inland species of sturgeon that gives a highly prized caviar.

subtle gradiations. Here the cline of the steppe is most visible. Many genera found throughout Mongolia have expressions and speciation in multiple subdivisions of steppe; for example, species of *Delphinium*, *Echinops*, *Caragana*, and *Astragalus* are distributed in them all. There is at present quite a bit of interest in—and concern for—the flora of Mongolia on a global scale, even as Mongolia is becoming more receptive to mineral development and mining becomes more common.

Plant-People Connection

As humans migrated north from the Tigris and Euphrates and crossed over the Caucasus Mountains, they found the great plains of Moldavia, a large steppe in which to hunt new species of animals and develop more stable foraging practices. Further migration was halted until two events occurred. At the time humans moved into this area, the Aral Sea connected to the Ural Mountains, making east-to-west movement almost impossible. Not until a period of drought led to the recession of the Aral Sea, in combination with the domestication of the horse, was the vast Central Asian steppe opened to broader migrations of humans. To this day, nomadic races exist in Central Asia. Many Mongolians still lead a migratory life, as do migrant tribes of the Kazakh and Tuvan people.

Humans have had a long and interesting relationship with plants in this part of the world, which has been opened to mainstream Westerners only over the last couple of decades. The Russians have been very active in the botanical study of these areas, and there are large volumes of information written in Russian about them. Botanical studies and horticultural pursuit are two very different arenas. To have seen a plant and collected a specimen is a wondrous thing. It is often the piece that drives the imagination of the horticultural collector and grower of

Horses on the Mongolian steppe.

plants. These specimens tucked away on dusty old shelves and their references in journals help to ignite the curiosity of the horticultural traveler and inspire the wanderlust of those who still carry in their DNA the heart of the nomad. Now, as populations of humans are moving inland and back to our grassland roots, there is a greater need for plants that can survive on limited water. Long-overlooked habitats like the steppe are hot-beds of diversity for gardeners.

In the middle of Kazakhstan is a vast desert that can shift seasonally into shortgrass steppe. The shift to shortgrass occurs only in years of higher rainfall or where there is persistent snow cover during the winter. The resulting corridor of grasslands is the basis and means of the Silk Road of legend. This corridor is also a superhighway for migrating birds that cannot cross the Tien Shan to get to their winter homes in India. The Chokpak ornithological station in South Kazakhstan catches, tags, and releases thousands of birds every year in order to track migration patterns. The west-to-east spring migration and the reverse east-to-west fall migratory pattern is a function of the direction of the Tien Shan mountains. The success of Chokpak is largely due to its location in a narrow valley, where the southernmost extension of the Karatow nearly abuts the Tien Shan. During the greater migration season, this area draws numerous species of raptors to capitalize on the concentration of migrating food.

It is interesting to postulate: did the first humans who ventured into these areas let the migration patterns of birds inform their travel routes through the great steppe and desert regions? Had they observed that certain birds could travel only a set distance before needing to find water? Or were they purely following migrating herds of larger ungulates, feasting upon them as they migrated? The Kazakh people have had a long tradition of utilizing birds to hunt game. This

A traditional Mongolian family home on the steppe.

Berkutchy, as it is called, is a lifelong bond between the hunter and his eagle, specifically the steppe eagle, *Aquila nipalensis*. However it happened, mankind found the steppe and turned it into the superhighway for travel and trade.

On the eastern slopes of the Altai is the Mongolian steppe, the stomping ground of the great Genghis Khan, to whom more than 10 percent of all Mongolians claim a direct lineage. It is the place most people think of when they hear the term "steppe," stirring up visions of hordes of warriors upon horseback, sacking ancient cities and driving their captives before them as human shields—all the while trampling dozens of *Astragalus* species under hoof and foot. Now this area is home to a more peaceful class of nomad. The culture and values of nomadic life are still very important to the cultures that exist within the borders of Mongolia, and it is more than unfortunate that much of that life is beginning to come to an end. Over the last decade, severe droughts and extreme winters have laid waste to millions of herd animals and forced many people to the capital, Ulaanbaatar. In the surrounding steppe, severe overgrazing coupled with drought is creating desertification of an already fragile steppe on a massive scale.

The Eurasian steppe is not close enough to any of the major oceans to have much of its climate influenced by them. There are some inland seas and large bodies of water that will help to distribute moisture along the steppes, but we humans have tapped into some of these areas, moving water around through irrigation canals to water-intensive crops like cotton, draining these areas to the point of excess and environmental catastrophe, as in the case of the Aral Sea. The Caspian Sea is still a major source of water vapor to be transported to the steppes; much of the west-to-east movement of the weather patterns are dictated by how moisture is formed in the Caspian and moved over the steppe by prevailing wind patterns.

OVERLEAF A stark contrast in present-day Ulaanbaatar, as traditionally nomadic Mongols are forced into the modern city that is springing up on the steppe.

Bactrian camels are dependent on a healthy steppe habitat.

Habitats and Plants

A grouping of plant life is contiguous all along the Central Asian steppe, but there are three really remarkable areas of transition, where the unique floras and the floristic feel of Central Asia blend with their neighboring floristics, repeating our Venn diagram on a massive scale. At the western end of this continuum is the great influence of the Mediterranean biome (including the Caucasus Mountains). On the southern end, with the Tien Shan separating the vast desert regions of the Taklamakan, the great Gobi, and the Junggar Basin, the influences of the Himalaya and more subtropical China are felt. Finally, to the north, the florisitics are influenced by the Altai and the pressures from Lake Baikal and Siberian taiga (maritime).

LAKE ZAYSAN

Lake Zaysan is on the eastern edge of Kazakhstan, very near the border with China. The main inlet and outlet is the Irtysh River. At one point Zaysan was quite shallow and broad, nestled in a low depression just north of the Tarbagatay range. Presently the Irtysh is being dammed, causing the lake to more than quadruple in volume. The Irtysh River is the major water source for Ust-Kamenogorsk and supplies the supporting croplands with irrigation. Very few moisture-bringing weather events happen here, and the area is ravaged by high winds for most of the year. The winds are so great here that to the east they have created miles and miles of sand dunes, which define the natural border between Kazakhstan and China.

On Zaysan's northeastern shore, in an area locally referred to as the Zaysan Desert, a hardened sedimentary cap overlays soft eroding gypsum clays, adding a beautiful rainbow effect to the landscape. This type of specialized geology is perfect habitat for a host of rare plants. *Rheum nanum* is one of the most fascinating, along with *Nanophyton erinaceum*, the trees of which are so twisted and gnarled they look to be hundreds of years old. This area is visually and geologically almost identical to the Vermillion Bluffs, in northwestern Colorado, and although neither *Rheum* nor *Nanophyton* occurs in Colorado, the few plants that do grow there look remarkably similar—for example, *Atriplex confertifolia* (shadscale saltbush), which also has trunks twisted and tortured by strong winds and eroding soils. The Zaysan Desert is also a near mirror of the Badlands in South Dakota. All three of these areas are dominated by large deposits of gypsum, creating badlands with very high pH. Only a few specialized plants can withstand these harsh carved canyon lands.

LAKE BAIKAL

Baikal, a UNESCO world heritage site, is believed to be a massive rift lake and the oldest lake of any kind in the world. It is reported that as much as a third of all the earth's freshwater is contained within it. Baikal also boasts having 80 percent endemism for life forms. The most famous is the endemic freshwater seal, *Pusa sibirica*. Baikal is entirely surrounded by mountains, which have helped this area escape evolutionary pressures for thousands of years. To the west are wet mountains thick with Siberian larch (*Larix sibirica*) and beautiful woodland plants like lady slippers and bearberries. To the east, however, a series of steppes leads to the vast grassland steppes of Mongolia. On this drier slope, springtime brings swaths of early pulsatillas in shades of lavender and yellow. Perhaps one of the most unique floral tales in this area is the blue skullcap, *Scutellaria baicalensis*, which is virtually identical to the North American *S. lateriflora*; the plants were used medicinally by both the indigenous shaman/medicine men of the Buryat Mongols of the Baikal area and the Cherokee, Comanche, and Apache tribes of the Great Plains. Which gives one pause—is the plant the same on both continents? Did the shaman bring the plant with them as they migrated? Or did they see something so similar that they simply assumed they were the same?

Starting from the western end of this area, there is a fairly stable continental climate with moisture coming mainly in the winter and spring, followed by long, hot, dry summers and cool fall weather patterns. As you move farther east the great Caspian and Black seas have less of an influence on precipitation; weather is

Nanophyton erinaceum, a rare plant found only in these high gypsum badlands.

*An emerging leaf on *Rheum nanum*.*

dictated by the Tien Shan and the Altai mountain ranges, and moisture events can happen at almost any time of the year, giving the feel of a monsoonal season during the summers and periods of extended drought during the winters. While snow and bitter cold are very common for the central portion of this region, the snow rarely stays on the ground long enough to soak in. Most precipitation that falls in the form of snow simply blows away in the strong and dominant winter winds.

As you move north of the steppe, a desert in the central portion of Kazakhstan is bordered on its north with another expression of steppe on the edge of the great boreal forest of Siberia. The size of the desert and bordering steppe in this interesting area is in a constant state of flux, as weather dictates their physical boundaries. In recent years the development of larger cities and exploitation of resources have given desertification the advantage. Astana, the capital of Kazakhstan, can see such temperature extremes as winter lows of −51°C and summer highs of 41°C, one of the more extreme shifts in the Asian steppe. Then to the east of the Altai mountains is the vast cold of the Mongolian steppe. Long, bitterly cold winters and short, hot, dry summers are the norm. Ulaanbaatar regularly sees average lows of −25°C in January and temperatures of 33°C in July and August as a record. The average growing season is only 90 to 110 days, and weather patterns typically move east to west. Along the borders of the Tien Shan, however, the direction from which weather systems originate dictates if they will be moisture-producing events. Storms moving from the southeast to the northwest will typically produce moisture events; east-to-west storms bring wind and clouds but little in the way of precipitation. If there is any moisture in these storms, it tends to be virga and of little relief to the dry plains.

The majority of the Eurasian steppe is not a natural fire cycle habitat, especially in the lands surrounding the Altai mountains. In the rare event that thunder and lightning are produced by storms, they are always associated with heavy rain events, preventing fire. This is worth noting, as many of the Eurasian species of plant that have become mobilized in the American West are very sensitive to fire. Controlled burning is the main control prescribed by many land managers to keep these overfriendly plants in check.

WESTERN EURASIA

The westernmost expression of steppe in Eurasia is very fascinating floristically. The area includes the Caucasus Mountains, north above the Caspian Sea and staying south of the Ural Mountains. This area boasts a 20 percent endemism in species; it is also the great grain belt, whence the Volga Germans hail, and home to many beloved garden plants, including the iris, peony, crocus, and campanula. This area can experience great temperature extremes from winter to summer, often having long and dry fall seasons. Most of the moisture for this region comes from winter moisture in the mountains and a spring rainy season. Hot and oppressively dry summers (July and August) are the norm here. Plants from this region have a distinct familiarity to them, as expressions of the European flora that has been the dominant feature in ornamental horticulture for the last five centuries or so. Here the Oncocyclus iris is the attraction and obsession of the plant collector, growing in such harsh conditions yet having such large and intricately beautiful flowers. In fact *Iris* is one of the genera that ties so many aspects of the boreal steppe together, from *I. missouriensis* in North America to *I. lactea* in Mongolia and Kazakhstan. There the speciation of *Iris* intensifies; as you move west toward the diversity center of the group, you encounter the Junos and the Arils in the driest steppe and deserts. Genera like this and *Daphne* keep plant collectors drooling and wanting to grow and recreate the habitat in which these gems thrive.

CENTRAL ASIA

Here, along the vast grassy plains that connect East to West, is possibly one of the most intrinsically beautiful regions of the world. All along the watersheds of the great Tien Shan, major cities have been built up and sacked repeatedly over thousands of years, as people have tried to bend the untamable steppe to their will. Even Stalin sought to control this area to no avail. The steppe is still too wild and unforgiving to be held in the grasp of only one culture. The nomadic heritage is so ingrained into the steppe that freedom and movement are paramount to the norms of the cultures here. Geopolitical boundaries shift on a regular basis in Central Asia, but the steppe remains. You may travel for days here and not see signs of civilization, nothing but a vast sea of grass woven together by an ever-changing palette of flowers. Bulbs are first to welcome spring, and species like *Tulipa greigii* make vast carpets of warm color. The long, cold, windswept days of winter have passed, and it is time to use preciously stored energy to send up flowers, set seed, and gather reserves for the hot, dry winds of summer. Anyone will tell you that to see a field of wild *T. greigii* in flower is an experience like no other; the size and scope of these wild fields each spring rival the tulip fields of Holland. Springtime on the steppe is a fleeting and glorious successional dance. Each group gets their chance to shine—after bulbs comes mints, then legumes and asters. All hold the stage for their respective season.

Tulipa greigii.

THE AKSU-ZHABAGLY NATURE RESERVE

All along the east-west path of the Tien Shan are several relict and smaller mountain ranges that add to the diversity of habitats and microclimates in this portion of the steppe. At the far western end of the Tien Shan, the jagged peaks soften to more of a rolling foothill habitat, still without forest cover; the two major woody

On the steep canyon walls of Aksu-Zhabagly is a forest of wild apples, *Malus sieversii*.

genera are *Malus* and *Juniperus*. In this part of the Tien Shan is Aksu-Zhabagly, one of the oldest and largest nature reserves in Central Asia and on the tentative list as a **UNESCO** world heritage site. Junipers dominate the woody plant community; looking out across the rolling shortgrass hills you see *J. semiglobosa* and *J. polycarpos* var. *seravschanica*, in contrasting forms dotted along the horizon like manicured cones and balls. In midsummer, when the river valleys are teeming with life, most of the low-lying steppe is already chasing into summer dormancy, and the skeletons of *Eremurus fuscus* and *E. turkestanicus* stand like candles waiting for strong winds to come along and blow their seed across the rolling hills, there to await fairer temperatures for germination.

In the canyons and river valleys of Aksu-Zhabagly, and particularly where the Zhabagly River leaves the reserve, are ancient forests of apples. This is the land of apples and the center of their distribution. For years researchers have been coming here to retap the original source of apples, assigning trees accession numbers and sending scion wood to Geneva, New York. The hope was to save a huge portion of that genetic diversity for breeding programs as well as to have it preserved in more than one site. *Malus sieversii*, the species from which all our farmed apples derive, contains a plethora of diversity. Virtually every tree presents a unique fruit, whether in texture, color, size, flavor, or storability, making this forest the "Garden of Eatin." The Zhabagly River that flows through this

Tulipa lemmersii,
gone to seed.

LEFT Bronze Age carvings
in the Aksu-Zhabagly
Nature Reserve.

valley is fed mainly by glacial melt and persistent snowfields, keeping this canyon particularly cool and allowing for the perfect microclimate for these apples to grow on the steep, rocky canyon walls. On the flat ridgetops, out of the cool canyon walls, grasslands stretch as far as the eye can see.

Another of the special places within the Aksu-Zhabagly is a mid-elevation montane steppe where there are a series of Bronze Age carvings on some flat reflective slabs of a black basalt. Growing in this area is the wonderful mat-forming *Pseudosedum affine* as well as massive populations of *Artemisia gmelinii*. Both plants are currently unknown in cultivation yet have incredible potential for western North American landscapes. The pseudosedum could and should have applications in not only perennial gardens but rock gardens and green roofs. The sage has a wonderful fragrance and a year-round presence, with large exposed foliate buds.

Life and the diversity of floristics change with nearly every new canyon or stream that breaks free of the great Tien Shan. The geology shifts so dramatically, it feels like every one of these isolated areas is teeming with its own set of endemic flowers, many of them tulips. One canyon in particular was incredibly interesting. Growing at different elevation bands throughout this eroded limestone canyon were plants like *Scutellaria immaculata*, *Acer tataricum* ssp. *semenovii*, and *Tulipa lemmersii* (a recently discovered species).

Continuing along the Silk Road toward Almaty, Kazakhstan's former capital, one encounters a vast sand dune area where the dominant vegetation is *Cannabis sativa*, a species native to this region (which is known locally as the medicinal forest). While it is considered acceptable to visit for some healing treatments, it is illegal to collect or transport plant material from here. During the flowering season, locals and visitors will camp in this desert for days or weeks, returning

Convolvulus tragacanthoides.

home only when they have had enough "medicine." Another interesting feature is that wild camels can often be seen in this desert, grazing, while this plant is nearing its flowering stage.

Almaty or Alma-Ata ("mother of apples"), nestled at the heart of the Tien Shan and the Dzungarian Alatau mountain ranges, makes for a great jumping-off point to visit some of the tallest mountains in the Tien Shan as well as really unique expressions of the steppe. Here you can see the greater influence of the Himalayan floristic pressure. Almaty was among the first cities in the former Soviet Union to run irrigation ditches specifically to promote tree growth. It does get a little more moisture in the form of rains, but the majority of water is runoff from the massive mountains that dominate every view in this, the largest and most populated city in Kazakhstan. From Almaty to the southwest near the Chinese border, there are some wonderful plants. Growing in some of the most harsh conditions is the choice and rare *Convolvulus tragacanthoides*, which carries its beautiful pink flowers on low spiny cushions. In some low areas, where there are seasonal streams and gullies, you find very showy stands of *Clematis orientalis*. Little else grows in these hot, dry, rocky areas, but the plants that do are wondrous and deserve to be cultivated.

Moving out onto the vast Mongolian Plateau, one gets a sense of just how massive and peaceful the steppe truly is. Here grasses truly do dominate the

The sun setting on the remarkable Ikh Nart Nature Reserve.

landscape visually. There are seas of grasses as far as the eyes and imagination can carry you. Tall sturdy specimens of *Stipa splendens* make their presence known in almost every habitat in Mongolia; known to the Kazakh and Tuvan people as chi grass, it is used as a fiber and can be woven into baskets or mats. The plants dot the steppe like wonderfully tall statues. But, to break your mind free of the monocotyledonous monocultural monotony, look beneath and between the lovely grasses. Surrounding them all are a wonderful array of forbs and wildflowers. Beautiful *Artemisia* species and some very interesting expressions of the Amaranthaceae like *Krascheninnikovia ceratoides*—with its wooly leaves and fuzzy-seeded inflorescence, it could be a mirror of Colorado's winterfat. You might also encounter *Acanthophyllum pungens*, an interesting subshrub only half a meter tall and wide, but with leaves modified into spines and clusters of deep pink flowers borne atop wiry stems. Continuing east past Ulaanbaatar the steppe continues, and as you draw closer to the Pacific, the moisture increases and the floristic pressure brings a new richness and host of endemic species, among them *Iris bungei* and *Lilium pumilum*. There is a vast untapped potential for horticultural exploration in Mongolia, and with its borders now open, there is also hope that we can bring some of these plant into gardens.

Alfredia cernua.

CENTRAL ASIAN STEPPE

ASTERACEAE

The aster family is vitally important to all steppe climates, truly a defining factor in many of the regions. The great sagebrush steppes are defined by and named for the aster family. Asters are also some of the most important agricultural crops; *Helianthus* is a major source of oil and food for humans. Horticulturally they are equally important; many are superb garden plants, with a nearly unmatched ability to attract pollinators and wildlife to the garden.

Alfredia cernua comes from the fringes of the montane steppe regions, all the way across Siberia into Kazakhstan and throughout the Altai. Inflorescence can be 4 to 6 feet tall and have nodding yellow flowers; large triangulate leaves comprise a basal rosette. Rarely cultivated in western gardens, it has wonderful potential as a large statement plant in the prairie garden. Another alfredia grows throughout the mountain ranges of Central Asia, but it tends to be a high-elevation plant that occurs only in alpine screes.

Echinops tschimganicus and other Eurasian *Echinops* species have made their way into some popular garden cultures. But just because they are often included in the English cottage garden, that doesn't mean that's what they like or even prefer. Along with the tall spherical flowers and the strange, almost glowing blue to white of the flowers is something else that makes a thistle garden-able and worthy. *Echinops ritro* has been the standard in the trade and is still used in many dryish cottage gardens. The real gems of the genus are *E. nanus* and *E. latifolius*. *Echinops nanus* is found more into the colder montane steppe of the Altai and the northern slopes of the Tien Shan; plants have a much more compact growth and are very densely covered in cobwebby fuzz on almost all parts, whereas *E. latifolius* is a much taller plant, with floral clusters a more vibrant blue. The foliage is still very much covered in the thick fuzz; however, the upper side of the leaves on *E. latifolius* are much greener. *Echinops latifolius* is abundant on the steppes that separate the Gobi Desert from the edges of Siberia. The prickly nature of the

Echinops tschimganicus near the Black Irtysh River, Kazakhstan.

Echinops nanus in Mongolia.

Erigeron krylovii near Lake Markakol, Kazakhstan.

RIGHT *Erigeron aurantiacus* with *Tulipa greigii*.

leaves and inflorescence make it unpalatable to livestock but extremely attractive to gardeners and plant enthusiasts.

Erigeron is common to all habitats of the Boreal Kingdom. From extreme elevations to the hottest and driest deserts, you will find some expression of the fleabane. The greatest number of species are found throughout North America. Eurasia is, however, a close second with some very interesting and beautiful expressions. *Erigeron krylovii* is quite rare in its native range. Plants occur only in dry steppes near Lake Markakol in Kazakhstan and along the edges of the Dzungarian Alatau and the Mongolian Altai and extending into the Gobi Altai ranges. Wonderfully fringed ray flowers glow a deep lavender with a contrasting orange-yellow eye.

Erigeron aurantiacus is among the few expressions of red or orange in the genus. Plant breeders have used this trait in early breeding experiments to add that color to the popular florist crop *Bellis perennis. Erigeron aurantiacus* is on its own a very adaptable perennial with a unique splash of color that fits very nicely into rock gardens, shortgrass prairie gardens, and the front of traditional perennial borders.

Inula orientalis is another of the great and bold sunflower family that is at home in the steppes of Central Asia. Large straplike leaves set below impressive 5- to 7-foot-tall multi-headed inflorescences, it is truly one of the floral world's best DYCs (damn yellow composites). The plants seem to be competing to be the tallest and most garish plant blooming in midsummer's splendor. This species sticks to the southern portions of the Eurasian steppe, growing in montane steppe and rocky valleys along the Tien Shan and extending its range into China.

BORAGINACEAE

The borage family is well known to herbalists and gardeners around the world. *Anchusa* and *Echium* are fixtures of both the Eurasian and South African steppes.

Echium amoenum, alpine unit, DBG.

Dianthus chinensis growing in shortgrass steppe, Ikh Nart Nature Reserve, Mongolia.

Dianthus hoeltzeri.

Echium vulgare is a prevalent feature in much of the grassland and true steppes all across Kazakhstan and Mongolia; the hairy and coarse leaves are rarely grazed upon, giving this plant the advantage over many of the grasses and forbs that are much tastier and less texturally offensive. Horticulturally one of the best performers in the genus is *E. amoenum*, a long-lived and very cold-tolerant species for Colorado landscapes. This plant is native to the southern steppes of Kazakhstan and can be found distributed among many of the relict mountain ranges extending from Turkey to Pakistan. Occasionally this plant has been confused with its close cousin *E. russicum*, which has dingy rust-colored flowers and is often biennial or short-lived. *Echium amoenum* is a much brighter red, fading to a slight rust. Promoted by the Plant Select program, it has been gaining regional popularity but still struggles outside of the steppe belt in North America. It tends to be intolerant of long periods of wet.

CARYOPHYLLACEAE

Dianthus chinensis grows in many different habitats in Russia, Mongolia, Kazakhstan, and China. Flower color changes (its former epithet was "versicolor"): blooms are either white or rich pink, depending on both daylength and temperature (I suspect this is to capitalize on various pollinators' being out and about at different times). Plants can bloom for a long time, starting in early spring until after the first frosts. The very adaptable *D. hoeltzeri* is native through

Sedum ewersii.

Pseudosedum affine.

Rhodiola semenovii near Big Almaty Lake.

Orostachys spinosa growing in desert steppe in Mongolia.

eastern Kazakhstan, southeastern Russia, Mongolia, and northeastern China, growing mostly in montane and shrub steppes; its lacy flowers closely resemble *D. superbus*.

CRASSULACEAE

The stonecrops of Crassulaceae serve many horticultural purposes in a variety of garden styles. Sedums especially are a staple genus of the green roof industry; they are typically easy to propagate from cutting and seed. Only about 8 to 10 inches tall, *Sedum ewersii* has great potential to be a wonderful garden plant: stems are rosy red, and the thick rounded leaves are very glaucous blue to almost white. In bloom, the plant shimmers with clusters of glistening pink flowers. Plants are mostly found on well-draining rocky outcroppings but have been spotted growing among grasses in shallow soils. *Pseudosedum affine* is a low-growing sedum, typically with olive-green stem and leaves, and white flowers in midsummer; it is found in rocky disturbed sites on the fringes of grassy areas. Visually and growthwise, it is very similar to *Sedum reflexum*, and it too would really perform well in rock gardens and on green roofs.

Another really interesting stonecrop is *Orostachys spinosa*, very common throughout Kazakhstan, Mongolia, and into China in the cold, dry steppe. It seems to be the evolutionary bridge between *Sempervivum* and *Sedum*: while

Orostachys spinosa blooming in the western foothill steppes of the Altai.

Ephedra equisetina, close-up of cones.

many of its features are like those of *Sempervivum*, it grows in places that a sedum would, and with a similar distribution. Small clumping plants set many offsets or pups, and as the rosettes mature, they flower and then die off. A more alpine stonecrop, *Rhodiola semenovii*, is endemic to the Altai mountains of Central Asia.

EPHEDRACEAE

The joint-fir family is one of the great connectors: most species of *Ephedra* have been used medicinally or shamanistically by indigenous people throughout the boreal world, and in the landscape they can play a major role, adding structure and texture. *Ephedra viridis* is the dominant feature in much of the Intermountain West, and *E. chilensis* crosses the great Andean range in a few places to creep toward the steppe from the coast. Neither have *E. equisetina*'s adaptability to garden culture or combination of ornamental features. *Ephedra equisetina* is, as *E. viridis* is in the West, the dominant feature in much of Central Asia. The forms in cultivation are much more robust than you often find in the wilderness. *Ephedra equisetina* grows as a large shrub and carries the most interesting blue hue all through the year. When the plant is in fruit, in good years, it is heavily adorned with bright red fleshy cones and almost appears to glow.

FABACEAE

The legumes of the world fill several crucial roles in natural ecological processes. Legumes are an incredibly important food source for all manner of life forms, including humans and other animals. Below the soil surface, legumes are associated with a root-colonizing bacterium that makes chemically-bound nitrogen available to plants. Once the host leguminous plant dies, the processed nitrogen

Astragalus species are among the various legumes of Central Asia.

Caragana microphylla.

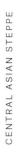

becomes a sort of slow-release fertilizer for other plants that share its rhizosphere. Clover-laced grass has long been a cover crop and green manure in the nursery industry for its nutrient-recycling ability and the way that the coarse roots of leguminous species break up heavy clays and allow for improved texture and porosity. Agriculturally speaking *Glycine max* (soybean) is a very important crop, cultivated in China for thousands of years; the United States and Brazil now lead the world in soybean production.

Astragalus, by some reports, lists 117 species in Mongolia proper alone. These plants will grow in every habitat from high alpine tundra to the deepest heart of the Gobi Desert; some of the most beautiful species are *A. junatovii* and *A. laguroides*, but two others are worth considering for the garden. *Astragalus mongholicus* has great potential as a smaller semi-woody subshrub or herbaceous perennial. Plants, which can be up to 3 feet tall and wide, have cream-yellow to white flowers on racemes with as many as 50 flowers per inflorescence; they add great texture to the steppe garden and attract a flurry of bees while in bloom. This species is very widely distributed throughout eastern Kazakhstan, Mongolia, and eastern Siberia, mainly growing in true steppe and the edges of forest steppe. More herbaceous, less woody, but about the same size, *A. onobrychis* also has great horticultural potential. Flowers are blue to pink, and the plant is more of a rosette. This species is very adaptable to a range of garden conditions and may even self-seed where it is happiest.

A couple of *Caragana* species have made it into production but have not yet been given the attention they deserve. *Caragana microphylla* is a dominant shrub on the steppes surrounding the Altai. This plant has enormous potential in the steppe garden, and some interesting cultivars are available. *Caragana microphylla* 'Tidy' is a selection introduced from the Cheyenne Research Station. It is mostly upright with a slightly open habit and an elongated finely dissected leaf. Stems are a golden color, adding to the winter interest of this plant. The late great plant-hunter Harlan Hamernik's selection 'Mongolian Silver Spires' is a very columnar form with strikingly silver leaves and clear, bright golden flowers—truly a stunning plant for the dry garden. Harlan also collected a pink-flowering form of *C. microphylla* on one of his many expeditions to Mongolia. I've not seen it but have heard from others who have. I still think about the perfect caragana, a columnar, pink-flowered, silver-leaved shrub that will survive to zone 2 and need no supplemental irrigation when established. Someday I will make it to where Harlan saw the pink peashrub, and I can continue his work, I hope.

IRIDACEAE

Iris lactea is found across the Central Asian steppes, most often in xeric soils and shortgrass steppes. Plants can grow at quite high elevations and seem to be mostly

Iris bucharica.

LEFT *Iris lactea.*

Ixiolirion tataricum.

unpalatable to grazing animals. They look almost identical to *I. missouriensis. Iris lactea* almost always blooms among the foliage; *I. missouriensis* mainly blooms above it. Coloration and habitat distribution is the same, however.

Iris bucharica, the most commonly produced Juno iris, performs in a wide range of garden conditions across the country but always preferring well-draining or rocky soils. These plants are spring ephemerals in most gardens. The foliage begins to grow in the late winter or early spring and is often one of the first bloomers in Colorado gardens, along with *I. reticulata* and crocuses. Distribution is across the central portion of Central Asia, mostly on sunny slopes of the western ends of the Tien Shan, frequently throughout Afghanistan, Uzbekistan, and Tajikistan.

IXIOLIRIACEAE

Ixiolirion is a monotypic genus that occurs only in Central Asia. Until recently ranked with the Amaryllidaceae, it is now in its own family, but the debate over how many species or subspecies there are rages on. The one that has been cultivated and makes a good garden plant is *I. tataricum*. Plants have blue to lavender flowers and grow best in rocky or well-drained steppe gardens. This plant naturalizes well in the rock garden or in a loamy dry perennial garden. Mixed into a naturalistic garden, the grassy leaves and long-lasting bloom help give a colorful complexity to the garden. Plants are superficially and visually very similar to the

Dracocephalum bungeanum.

North American genus *Triteleia*, which includes some blue-flowering species that grow in similar habitats. *Triteleia ixioides* is very similar morphologically; the difference is the yellow flowers with a brown stripe rather than the blue of *Ixiolirion*.

LAMIACEAE

The mints of the world are a very large family, now boasting (thanks to recent phylogenetic studies) around 235 genera, encompassing everything from trees to high alpine cushions. True mints have been used as teas and medicines for centuries. Mints like oregano, winter savory, hyssop, rosemary, and thyme have been used culinarily for just as long. The Kazakh and Tuvan people use *Ziziphora* species to flavor their kumis (fermented mare's milk); this gives it that minty fresh taste that you can only expect from the finest Central Asian kumis.

The literal translation of *Dracocephalum* is "dragon head," and when you look at the flowers of any of the 60-plus species in this genus it is easy to tell where they came up with that generic name. Two very little-known species deserve a deeper horticultural review. *Dracocephalum bungeanum* grows in true steppe and montane steppe in Mongolia. Plants are adapted to dry, rocky, shifting soils and thrive where there is little protection or competition. In areas that are heavily grazed, it hugs low to the ground and finds respite in rocky fields or talus. Small silvery scalloped leaves are surmounted by dense clusters of purple bracts with blue flowers protruding. *Dracocephalum nutans* is a much taller plant with a more elongate inflorescence but still shares the dark bracts offset by the blue flowers.

Nepeta sibirica.

LEFT *Dracocephalum nutans.*

Both plants are extremely adaptable to a very wide range of conditions, making them useful in the garden.

Nepeta sibirica, a large (up to 3 feet tall) and very long-lived perennial with large spikes of blue flowers, is often found in montane steppe in the Tien Shan and the Altai mountains. In Siberia proper, it is more common in lower meadows. Dark green leaves are long and serrated, on sturdy stems. Plants can take a wide range of conditions; their ultimate height is greatly influenced by how much water they receive.

Phlomis cashmeriana is a wonderful perennial that comes from the western end of the Himalayan complex and the dry high-mountain steppes of Kashmir. The Plant Select program recommends it as a plant that will thrive in cold windswept steppe gardens along the Front Range of Colorado. Long, fuzzy strap leaves make an attractive rosette, and the 30-inch verticillasters that adorn the plant in midsummer stand tall well into winter, giving structure and texture to the winter landscape. Little known in the western world, *P. oreophila* is about half the size of *P. cashmeriana* and even more drought-adapted; smaller leaves still make a very attractive rosette, and the verticillasters are borne upon almost black stems. Last time I was in Kazakhstan I encountered a species that grows in salt wastes and on the fringes of dunes in very poor steppe. It was *P. salicifolia*, a sprawling rambler that even in terrible conditions forms a plant 4 to 6 feet in diameter. Like the aforementioned species, it bears lavender flowers, but at the base of each verticillaster are two long and very sharp spines. As its epithet suggests, the foliage is long and narrow, very lanceolate.

Salvia is easily the largest genus in the family, and with nearly 1000 species that have been used medicinally, culinarily, and horticulturally, it is very unlikely that anyone has not encountered it in one form or another. *Salvia deserta* is

Phlomis oreophila in Mongolia.

Phlomis salicifolia.

Salvia caespitosa.

RIGHT *Salvia deserta.*

Scutellaria cordifrons.

the dominant species in all types of Central Asian steppe. Visually it resembles *S. nemorosa*, only distinctly larger and more robust. Unfortunately *S. deserta* has the potential to escape cultivation. Flowers are typically blue to dark purple, but there are soft pink forms as well as albinos.

Salvia caespitosa is native to the western portions of the Central Asian steppe along the Anatolian Plateau. It is a wonderful groundcover, featured in rock gardens for quite some time yet surprisingly underappreciated in the broader gardening culture. It forms low mats, covered with white flowers perched atop. The size of the individual flowers as compared to the foliage is something unexpected and unique in the genus. Plants grow quickly in sandy soils or on gentle inclines, blooming in late May and early June. I have seen specimens grown well on and in drystacked walls. Their hairy foliage makes them susceptible to rot and melting-out if given too much overhead irrigation.

Scutellaria cordifrons is a low-growing subshrub found throughout the greater Tien Shan range. Flowers are yellow, often purple on the tops of the hoods. They almost resemble a clutch of snakes emerging from a shrub. The triangulate scalloped foliage is fuzzy and kind of sticky, with a definite minty fragrance. Found in rocky disturbed areas, this plant is a natural for the rock garden. *Scutellaria*

baicalensis (named for Lake Baikal) is regularly cultivated in ornamental gardens as well as having a long history in medicinal uses. This long-lived perennial grows a thick root for energy reserves to survive long periods of drought as well as bitter cold winters. Its rich blue flowers have a white stripe down the center of the hood (hence skullcap, one of the generic common names).

LILIACEAE

The lily family was once among the largest in the monocots, but with new advances in genetics and recent phylogenetic studies, it too has been greatly reduced. Three very important genera to the boreal steppe regions are still contained within it, however: *Lilium*, *Tulipa*, and *Calochortus*. Tulips are a major food source for the wild Eurasian boar. The boar will actually cultivate tulip gardens in the montane steppe, using its tusks to dig up bulbs for food, but here's the interesting part: the boar leaves the seedlings or bulbs that are not mature enough to flower. The young bulbs then have a well-aerated plot, with no competition from surrounding plant material, and can establish and grow strong. The boars return to the same cultivated beds to feed approximately every three years.

Tulipa greigii, often considered the king of tulips, is where much of the large Dutch hybrids have their parentage; in the wild, there are typically only reds and yellows. *Tulipa turkestanica* is more of an alpine steppe species. The long,

Tulipa greigii.

Tulipa kolbintsevii.

glaucous foliage is attractive but the predominantly white flowers with a yellow eye is a real treat. Inflorescence can have as many as 12 flowers per stalk, which makes this species an important component of many breeding programs, as the majority of species tulip are single-flowered.

Tulipa kolbintsevii, a plant from the northern Tien Shan first described in 2012, proves that there is still a wonderous world waiting to be discovered on the Central Asian steppe. This species has three distinct and separate petals, white at the edges and rich golden yellow at the center; it is unique that the sepals are also white and yellow. The majority of tulips have three fused petals and hidden or absent sepals.

MALVACEAE

Alcea rugosa (Russian hollyhock) is a wonderful, very long-lived perennial for the garden, blooming reliably for well over a month. Plants have multiple flower stalks, with flowers mostly in yellows or creams. They are excellent pollinator-attractants and are much more resistant to hollyhock rust than is *A. rosea*.

PAEONIACEAE

The peony family is monogeneric, containing only true peonies. Neither of the two *Paeonia* species in North America is in broad cultivation. The rest are distributed throughout Eurasia. The garden peony has long been a treasure, and the tree peony has been cultivated for millennia in China and the Far East. Some very

Alcea rugosa.

Paeonia tenuifolia.

RIGHT *Papaver nudicaule*, just one of the poppies from montane steppe in the Altai.

Glaucium fimbrilligerum.

choice species for the rock garden or for the specialty gardener or collector are native throughout the Mediterranean region; however, at least one species that thrives in the steppes of Central Asia is worth mention.

Paeonia tenuifolia can be found growing in the foothills of the Altai mountains. I first encountered this wonderful perennial, with deep crimson flowers as much as 6 to 8 inches across, in a mixed-grass prairie with *Clematis integrifolia*, *Daphne altaica*, and *Rosa spinosissima*—truly a wild garden to compete with the best-laid plans of men. The dissected foliage gives it the common name fernleaf peony. It has been grown by a few speciality nurseries but has never been mass produced. Even rarer to find in the trade is the double-flowered form, which is the prize for many rock gardeners. *Paeonia tenuifolia* can become floppy if it is grown in too much shade or in too rich a soil.

PAPAVERACEAE

The poppy family is a standard for almost any garden, and poppies are native throughout the Boreal Kingdom. The most famous of all is *Papaver somniferum*, cultivated for centuries not only for its medicinal properties but for its ornamental beauty. Most poppies, including *P. nudicaule*, are found as montane or alpine plants; several can be found as meadow plants, even steppe-growing annuals.

Glaucium is restricted in its distribution to true steppe and desert steppe. *Glaucium fimbrilligerum* is one of the smallest in the genus but has great potential

Veronica liwanensis.

for the dry rock garden. *Glaucium grandiflorum*, as its epithet implies, is a wonderful large-flowered plant with bright orange flowers. Plants flower at about 2 feet tall and continue to bloom for several weeks, and the strappy, hirsute, and undulate rosettes are as attractive as the flowers as an evergreen statement in the garden. A much smaller plant but equally beautiful is *G. acutidentatum*; it has a dark ruddy orange-red flower and has even been seen with a bicolor form, dark in the center and golden orange toward the outer rim of the petals. None of the aforementioned species has shown any propensity to escape cultivation, but *G. corniculatum* has been put on the watch list in several states for invasiveness.

PLANTAGINACEAE

Veronica liwanensis, one of the first Plant Select introductions, comes from montane steppe in the Caucasus Mountains of Turkey and Georgia, and upon its introduction, the face and vision of steppe gardening was changed forever. With glossy green persistent foliage and shocking electric-blue flowers, this species is a must-have mass planting for gardeners facing water restrictions. It is certainly among the most important Eurasian steppe plants to find a home in the domestic horticultural marketplace. *Veronica pectinata* is a delightful wooly groundcover, native to the southern portion of the Black Sea; this plant is wonderfully adapted to the Colorado steppe: the dry winters along the Front Range help keep it from rotting when it is dormant.

Limonium binervosum.

Large hummock of
Acantholimon alatavicum
in alpine steppe, Pakistan.

Goniolimon tataricum,
variegated form.

RIGHT *Bukiniczia cabulica.*

PLUMBAGINACEAE

Acantholimon is one of the genera that helps to define the steppes. The distribution of the nearly 300 species stretches from the Caucasus and Moldova all the way through Mongolia. Species can be found north into Russia and south to the Arabian Peninsula. Flowers range from white to dark pink, almost red. They need mineral-rich soils and rocky disturbed sites. *Acantholimon albertii* grows on the northern slopes of the Tien Shan. Plants grow in true steppe at the far western edge of the Aksu-Zhabagly Nature Reserve and north into the Karatow. Small hummocks can be found growing among rocks and along disturbed sites. Another montane steppe species, *A. acerosum*, will come out of the mountains and grow in desert steppe; it is native to Turkey and the Caucasus. This is one of the first species to be grown at Denver Botanic Gardens. Plants form respectable hummocks in a relatively short time. Flowers are a rich dark pink; the seedheads hold on well into the fall. *Acantholimon alatavicum*, the presently accepted name for what was once several species, is native throughout the Tien Shan in Kazakhstan, Uzbekistan, Tajikistan, China, and Pakistan. Plant can reach 5 feet across and 2 feet tall. These prickly thrifts are wonderful garden plants. They can be slow to establish, and a little painful to work with at times, but the flower display is well worth it. That sounds like a description of me; maybe that is why I am so enamoured of this genus.

Goniolimon is a very interesting genus. The natural distribution of *G. tataricum* (Tatarian statice) is all along the Central Asian steppe through Russia and Kazakhstan and well into Mongolia, as its epithet indicates. It is a short rosette-forming perennial that has a relatively large inflorescence compared to the size of the plant. Plants bloom in late summer and leave a structural, almost tumbleweed-like stalk that adds a lot of texture to the dry garden. Very similar

Stipa pennata.

to *G. tataricum* but much smaller is *Limonium gmelinii*, which is distributed through true and desert steppe in Mongolia and in the southern parts of the Tien Shan into Pakistan and Afghanistan. Both plants deserve a closer look for cultivation in steppe gardens, as do *Acanthophyllum pungens* and *Limonium binervosum*, two other little-known species. *Bukiniczia cabulica* looks and blooms like *Goniolimon* except for the very intricate markings on the wide leaves, which makes the entire rosette look like a lace doily; these plants are monocarpic and are truly best enjoyed while still in the rosette stage.

POACEAE

Stipa splendens (chi grass) is a large clumping grass. It is uncommon to find many plants growing together. Solitary plants are a rich dark green with stiff upright foliage, to about 3 feet tall; when in flower, plants can reach 6 feet. They are very tolerant of high-alkaline soils and can be used as a forage crop in very salty fields. Many nomadic peoples weave this grass into baskets, mats, hats, and other rough textiles. *Stipa pennata* (feather grass) is a wonderfully graceful perennial bunching grass with very soft seedheads. This grass should be grown for the sense of movement it adds to a garden—that feature alone gives the plant a very long season of interest.

Atraphaxis spinosa,
close-up of seeds.

RIGHT *Rheum compactum*
in Mongolia.

POLYGONACEAE

The polygonums are a very interesting and adaptable family, with several great genera and many yet-to-be-cultivated species. The beauty and diversity of the North American *Eriogonum* is something to be experienced in the wild; *Polygonum* species of Central Asia are no less intriguing or diverse.

Rheum is a genus that has been cultivated for centuries, but some of the most interesting steppe species have been all but ignored. *Rheum tataricum* grows in sparse gravelly steppe throughout Kazakhstan and into Mongolia and China. This plant takes advantage of the short window of spring moisture to leaf out and send up a 4- to 5-foot-tall branched inflorescence. Two of the smallest species, *R. nanum* and *R. compactum*, also have great potential as garden plants. *Rheum nanum* is the smallest of the genus and grows in almost desert conditions; I first encountered it growing on a high cliff overlooking the Zaysan Desert, growing above one of the only sites of *Nanophyton erinaceum* in Kazakhstan. *Rheum compactum* is just a little larger but grows in a much different habitat, at much higher elevations, where it is subject to colder temperatures and the potential of snow cover for the winter.

Atraphaxis spinosa is a common shrub growing all along the northern borders of the Tien Shan and out into the steppe. Plants are usually about 3 feet tall and have a mostly rounded open habit, with very small glaucous leaves and insignificant white flowers on slender branches. What truly is outstanding and beautiful about this species is its fruiting stage, with the color of its papery seeds varying from white or soft green all the way to almost ruby-red. Plants have seeded

Clematis integrifolia 'Mongolian Bells'.   Clematis hexapetala.

around the bases in trials but not moved very far. More evaluation is needed before we can confidently call this a well-behaved garden plant.

RANUNCULACEAE

Clematis integrifolia is a wonderfully adaptable perennial. It can be found growing all around the Altai in the surrounding steppes and in the foothills. One of the upright herbaceous forms, this species mixes very well with grasses or in classic perennial borders. Typically its flowers are blue; however, Harlan Hamernik found a population with a wide variety of colors—everything from white to rosy pink to lavender to blue—and this is the form that Plant Select promotes as 'Mongolian Bells'. It is very similar in habitat, distribution, and look to *C. fremontii*, a wonderful North American native; however, *C. fremontii* is a bit slower to produce and therefore not commonly available commercially. Neither vine, shrub, nor herbaceous perennial, *C. hexapetala* occupies a strange niche. Its long arching stems are graceful on their own but when covered with pure white flowers—all facing up as if they were tied to the stems as ornaments—the plant is very striking. *Clematis hexapetala* comes from true steppe in Mongolia, as far west as its borders with Siberia.

Delphinium elatum is one of the original parents of the tall cottage garden delphiniums. I first saw this plant in low true steppe just to the west of the Black Irtysh River, before the foothills of the Altai, where an accidental fire had burned a low swell. Maybe only 30 acres were affected, but you could follow the path of the burn by the enormous swath of electric-blue delphiniums, standing near

Delphinium elatum in a recently burned field near the Black Irtysh River.

Rosa platyacantha.

6 feet tall (with no staking!). Most delphiniums have a chemical dormancy in the seed, making them difficult to germinate. It seems that this fire, and the way the soil chemistry changed, triggered a massive germination event. One of the only yellow delphiniums, *D. semibarbatum*, grows out on the shortgrass steppe; interestingly enough this species is also found on the Iberian Peninsula, but the majority of the plant populations are from the Central Asian Kazakh steppe.

ROSACEAE

Rosa albertii is native to western China, Kazakhstan, Mongolia, and Siberia. Plants make a fairly sizable shrub, to about 6 feet, and are most often found growing on the edges of forest steppe at 7500- to 9000-foot elevations. Flowers are almost always white to soft yellow. *Rosa platyacantha* comes from a slightly lower elevation band; in lower latitudes it grows closer to true steppe and shrub steppe. Plants are most often a large shrub, with large, richly fragrant, single

Spiraeanthus schrenkianus.

Pyrus regelii.

Dictamnus angustifolius.

yellow flowers. The dark glossy green foliage is very resistant to leaf spot diseases. Beyond the attractive foliage and large flowers, possibly the best feature is the dark red hips. The fleshy fruits are very abundant late in the season and hold on long into the winter, a good source of forage for birds as well as a continuing source of visual interest.

Spiraeanthus schrenkianus is a very rare Kazakh endemic, in exposed and windy true steppe and desert steppe; I witnessed it growing in very dry desert steppe in the northern rain shadow of the Tien Shan, west of Taraz in the rolling hills that lead into the heart of the Karatow. Plants make a small to medium-sized shrub, about 3 feet tall and twice as wide, with mostly white to very blush-pink flowers.

Pyrus regelii is a unique cutleaf pear that bears small yellow fruits that are very sweet but maintain that gritty texture. Plants are 6 to 10 feet tall and wide. The foliage is very dissected and has a fine texture. I encountered this plant growing in a very interesting valley in southeastern Kazakhstan, where *Tulipa greigii* yellow and red forms meet and hybridize, making wonderful bicolor flowers; many other wonderful species grow in this valley, including *Fraxinus sogdiana* and *Malus sieversii*, the wild ancestor of our apples.

RUTACEAE

Dictamnus angustifolius is restricted to the Altai foothills, where it grows with *Rheum compactum*, *Hypericum perforatum*, and *Daphne altaica*. Plants are smaller than most other members of the genus. Flowers are white and streaked with red venation; the foliage is very glossy and really stands out.

THYMELAEACEAE

Among the most sought-after alpine and rock garden plants are members of *Daphne*. The majority of the speciation is in the western Eurasian mountains—the

Diarthron altaica.

Daphne altaica.

Caucasus, the Dolomites, and the Urals. A few species come into the eastern Asian steppe and up into Siberia. There are even a couple of relatives in North America; these, however, are found only in the far northern parts of the United States and into Canada. *Daphne altaica* is among the largest daphnes, up to 6 feet tall and wide; plants are fairly common at mid-elevations in the Altai in shrub steppe, with clusters of white flowers that bloom for a very long season. These wonderful shrubs have yet to be cultivated in the United States.

Diarthron altaica is one of the few herbaceous members of the family. I encountered this plant in a montane valley in the low Altai of Kazakhstan below Burkhat, the highest peak on the Russian side of the border. This plant has tight clusters of tubular flowers that are dark red-pink on the outside, opening to white. Plants do not set seed very often; when they do, clusters of reddish brown drupes are formed. Leaves are an almost succulent glaucous blue, and plants are upright even with heavy flowerheads.

XANTHORRHOEACEAE

Eremurus, formerly a member of the Liliaceae, is now included in Xanthorrhoeaceae; this makes sense, as plants are corms, not true bulbs. Cultivars include selections in white, pink, and—probably the most stunning—the orange

Eremurus tianschanicus with *Inula magnifica*.

Eremurus fuscus.

'Cleopatra'. *Eremurus* species are some of the most interesting horicultural introductions of the last quarter-century. Every June at Denver Botanic Gardens, people flock to see the mass plantings of *E. stenophyllus*, easily the predominant species in the marketplace. Only a very few species and hybrids are commercially available, which is a pity because many wonderful species occur throughout the Central Asian steppes.

Eremurus tianschanicus is possibly the largest of them all. I encountered this plant near Almaty in montane steppe, growing outside of an old Russian cemetery on a rocky hillside with *Lavatera thuringiaca* and *Inula magnifica*. Flower stalks in this population were well over 12 feet tall. Individual blooms are white with a pink vein, giving the whole flower stalk a soft pink glow. Both *E. altaicus* and *E. fuscus* are smaller in flower and on the yellow to almost brown end of the spectrum; plants are only 2 to 3 feet tall and have a slightly more open flower stalk. Both species hail from the dry foothills and are very drought tolerant.

Eremurus 'Cleopatra'.

THE CENTRAL NORTH AMERICAN STEPPE

LARRY G. VICKERMAN

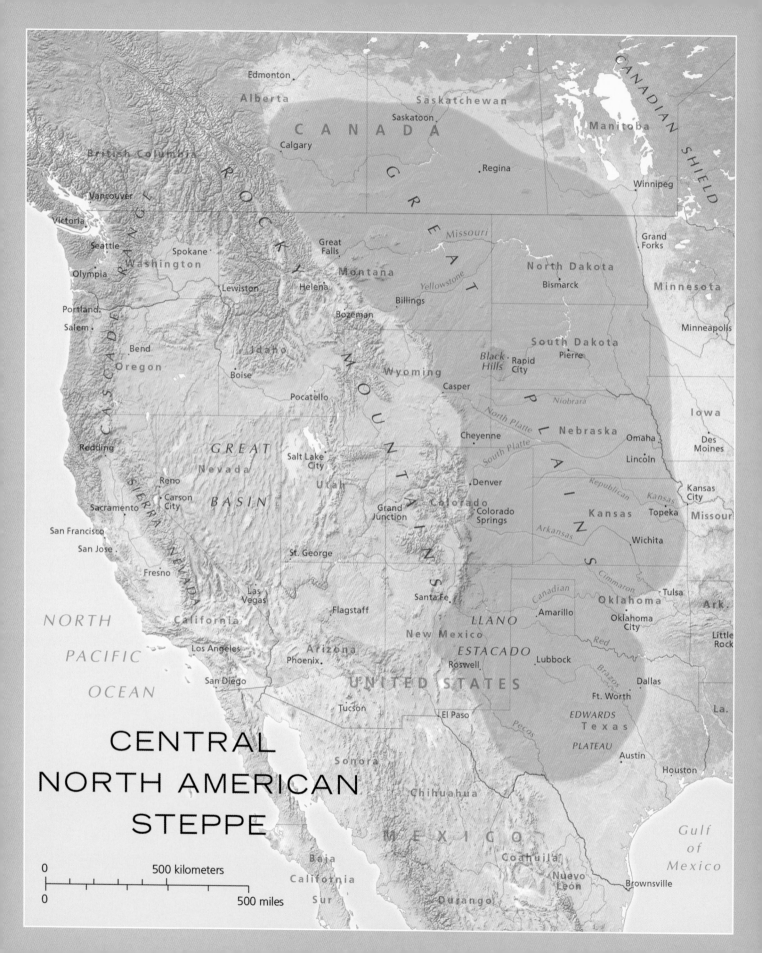

CENTRAL
NORTH AMERICAN
STEPPE

NORTH

PACIFIC

OCEAN

Gulf
of
Mexico

0 ————— 500 kilometers
0 ————— 500 miles

British Columbia

Edmonton
Alberta   Saskatchewan
                 Saskatoon
Calgary   Regina

CANADA

CANADIAN SHIELD

Manitoba
                           Winnipeg

Vancouver
Victoria
Seattle
Olympia
Washington
Spokane
Portland
Salem
Oregon
Bend
Redding

San Francisco
San Jose
Sacramento
Fresno

Los Angeles
San Diego

Reno
Carson
City

Las
Vegas
California

Tucson

Phoenix
Arizona

Sonora

Baja
California
Sur

Chihuahua

MEXICO

Durango

Coahuila

Nuevo
León

Brownsville

CASCADE RANGE
SIERRA NEVADA

Lewiston

Boise

Idaho

Pocatello

Helena
Bozeman
Billings

Montana

Great
Falls

Missouri

Yellowstone

ROCKY

MOUNTAINS

Wyoming

Casper

Cheyenne

North Platte

South Platte

Denver

Colorado
Springs
Colorado

Santa Fe

New Mexico

LLANO

ESTACADO

Roswell

El Paso

Pecos

GREAT

BASIN

Nevada

Salt Lake
City

Utah

Grand
Junction

St. George

Flagstaff

UNITED STATES

North Dakota

Bismarck

South Dakota

Black
Hills

Rapid
City

Pierre

Minnesota

Minneapolis

Grand
Forks

Iowa

Des
Moines

Nebraska

Omaha

Lincoln

Niobrara

Republican

Kansas

Topeka

Kansas
City

Missour

Wichita

Arkansas

Cimmaron

Oklahoma

Oklahoma
City

Tulsa

Ark.

Little
Rock

Amarillo

Lubbock

Canadian

Red

Brazos

EDWARDS

PLATEAU

Texas

Ft. Worth

Dallas

La.

Austin

Houston

GREAT

P

L

A

I

N

S

AMERICANS TEND TO think of the central North American steppe as a boring and featureless void separating the eastern woodlands from the Rockies and the recreational playgrounds of the mountainous west. I had the opportunity to spend ten years living there, working for Dyck Arboretum of the Plains just north of Wichita, Kansas, and I developed a very extensive knowledge and a deep appreciation of the steppe, with its rolling to flat topography, endless skies, frequently terrifying weather extremes, and amazing array of plants. Subtle changes in topography—punctuated occasionally with dramatic buttes and canyons layered with constantly changing soil substrates—create breathtaking niches and vistas filled with resilient and beautiful flora. Granted, the steppe has been heavily farmed and grazed over a large portion of its area, giving rise to huge swaths that display little to none of their historic plant or animal diversity. What you can find, if you are willing to take the time to look, is the remnants of an amazingly rich ecosystem that supported massive populations of animal life and in turn harbored complex and varied human nomadic cultures. Today, the steppe landscape has been tamed and fragmented, and the large wild herds vanished long ago to be replaced by cattle and commodity crops; but much of the plant life endures in fragmented patches, both large and small, scattered across this great prairie canvas.

One hundred million years ago, much of the area was an inland sea. Seventy to 80 million years ago, during the mountain-building period of the Laramide orogeny, the progenitors to the Rocky Mountains were thrust upward, helping to drain the inland sea. Millions of years of subsequent glaciation and erosion created a huge outwash of debris eastward, creating deep soils over the limestone and sandstone substrate as successions of shallow inland seas formed over the area through the millennia. Even today it is easy to find sandstone and especially

Niobrara chalk monoliths, Gove County, west-central Kansas.

limestone outcroppings across the region loaded with fossilized marine shells. In far western Kansas, the Chalk Pyramids, a series of Niobrara chalk formations capped by hard stone, rise up like Stonehenge on the flat plains. Farther east, the Mushroom Rocks feature sandstone concretions that have resisted erosion; these formed during the Cretaceous period, along the edge of the prehistoric sea. The Post Rock region of north-central Kansas has unique yellow-hued limestone with thick, even layers that were quarried for fence posts and other building materials in the treeless areas of the prairie. The Red Hills of south-central Kansas feature dramatic sandstone bluffs and mesas littered with gypsum crystals. In Clark County, Kansas, the St. Jacobs Well and Big Basin area is a mile-wide limestone sinkhole bordered on the west by sandstone, with accompanying changes in vegetation. Dramatic uplifts of Dakota sandstone appear at the Flatirons near Boulder and elsewhere along the Front Range of Colorado.

The rivers of the steppes, mostly arising in the Rocky Mountains, contributed greatly to the present landscape. The mighty Missouri arises in southwest Montana and makes a broad loop up through the Northern Plains before plunging south and east, eventually collecting every source of water from Palmer Ridge northward, with tributaries draining from two Canadian provinces and ten states. The Platte, with its braided and erratic flows, has moved billions of tons of sand and silt into the plains and helped provide material for the building of

RIGHT Fossilized marine shells in limestone from the Smoky Hills of north-central Kansas.

BELOW The Flatirons near Boulder, Colorado, are Dakota sandstone remnants of a massive uplift during the Laramide orogeny.

the Sand Hills. To the south, the Arkansas, besides being a lifeline through a very arid section of the plains, has also contributed water to the Ogallala Aquifer and moved sand to build some unique and isolated sand prairies, especially along the southern edge of the Great Bend. The Pecos, Cimarron, and Canadian rivers to the south have all carved significant drainages throughout the dry high plains, creating significant refugia for a host of woody and herbaceous plants.

Geography and Climate

The central North American steppe spreads eastward from the foothills of the Rocky Mountains to the more humid regions of the Midwest, bordered on the north by the Aspen Parklands of the Canadian prairie provinces and to the south by the eroded and striated Edwards Plateau of central Texas. Ranging 300 to 500 miles wide and 1800 miles long north to south, the steppe is a vast tableland tilted downward to the east and south. To the east, the mixed hardwood forest becomes more prominent in patches as you approach the 95th meridian, and rainfall increases to 32 to 36 inches of precipitation annually. To define the edge of the steppe on the eastern boundary is not easy. The balance between forest and grassland in these sub-humid regions is constantly in flux, based on localized geography, fire regimes, and human disturbance/agricultural activity. The Flint Hills of Kansas lend a clear-cut delineation; from this point, south and west in Oklahoma, tree cover tends to increase well out toward the 98th meridian. Beyond, below the Oklahoma Panhandle, the vegetation resembles shrub steppe, dominated by sand shinnery, mesquite, and occasional patches of sumac, until you reach the drier areas of the Texas Panhandle, where shrub growth is restricted to a few river valleys traversing west into New Mexico.

Farther south the dry shortgrass of west Texas gradually transitions into the Chihuahua Desert. Farther west, the towering Front Range of the Rockies breaks down in New Mexico and Texas into a series of mountainous sections that make delineation between steppe and true desert to the southwest ambiguous. In the northern range of the steppe, open grassland historically extended into western Minnesota and Iowa, with heavier forested sections like the lower Platte River in Nebraska jutting westward for several hundred miles out to Lincoln. From the west, intermittent peninsulas of conifer forest, including the Black Hills, stretch out from the Front Range to mingle with the sea of grass.

The Rockies, which reach their pinnacles in Colorado at over 14,000 feet, are a major influence climatically for this region, a formidable barrier to moisture-laden storms driven by the jet stream from the Pacific Ocean; don't forget, however, that the High Sierra, the Cascades, and the Wasatch Front have each already wrung moisture from these storms, leaving even less to fall in the rain

Spring storm clouds form over the warming plains.

shadow, on the lee side of the Rockies. To the south the Rockies recede into lower scattered ranges across the Chihuahua Desert, trailing south from the towering Sangre de Cristo and grading into the Guadalupe Range, with heights of 6000 to nearly 9000 feet.

Air masses affecting the region can come from all directions, depending on the season. In winter the prevailing airflow is from the west, creating downslope winds as they crest the Rockies and descend into the plains. Sunny days, occasional strong winds, low precipitation, and minimal humidity rule. Also in winter, depressions in the jet stream can sink southward, allowing cold arctic air masses to plunge out of Canada with no major landforms to deflect their ferocity—hence an old saying in the region: "There is nothing between the Arctic and the Great Plains but three strands of barbwire." Cold driving winds with minimal precipitation can lower temperatures dramatically in a few hours. Wild fluctuations in temperature are the norm: lows in the northern steppe can reach −40°C, −12°C in the southern areas. Conversely, high temperatures range from 35°C in the north to 46°C in the south (Lauenroth and Burke 2008).

In March, April, and May, low pressure systems moving across Arizona and New Mexico can create upslope. The counterclockwise rotation of low pressure systems can gather warmer, moist air from the Gulf of Mexico and push it up against the Front Range of the Rockies, bringing some of the heaviest snows of the season up and down the High Plains. As these lows move east, they continue to pick up moister air from the Gulf, intensifying the precipitation all the way across the steppe.

As the ground heats up in spring, cooler jet stream air aloft can cap the warm air rising from the surface. Eventually the buoyant warm air punches up through the cap. Fueled by warm moist air from the Gulf, the instability produces short, intense periods of rain, hail, gust fronts, violent thunderstorms, and localized tornadoes, especially from April through July. The central portion of the steppe is known as Tornado Alley because intense updrafts can turn these horizontal rotating air masses vertical.

Progressing into summer, the jet stream moves northward, and weaker low pressure system conditions abate. Thunderstorms produce intense rain over more localized areas, for shorter periods of time. Moisture coming from the southwest out of the Pacific Ocean and up from the Gulf of Mexico can build into heavy rains, creating the North American Monsoon. By fall, the monsoon flow weakens, and stationary high pressure creates long stretches of brilliant, dry, sunny weather. As the jet stream begins to move southward again, some days of cool, wet weather will occur. Snow can fall as early as September, primarily in the Northern Plains.

Precipitation follows a west-to-east gradient that can be as low as 8 inches annually directly in the shadow of the Rocky Mountains, increasing in north-to-south bands up to 32 to 36 inches annually moving east, as you tap into the more humid air flowing up out of the Gulf. Over the majority of the steppe, spring and summer precipitation accounts for 75 to 95 percent of the total moisture, depending on elevation and latitude (Lauenroth and Burke 2008).

The 100th meridian is a defining line in the plains, roughly following the 2000-foot elevation line across the steppe. This line of longitude follows the western border of Oklahoma (minus the panhandle) and generally splits Texas, Kansas, Nebraska, and the Dakotas, trailing northward to form the border of Saskatchewan and Manitoba. West of the 100th meridian, rainfall is generally less than 20 inches annually. Moving east from the meridian, precipitation increases along fairly distinct bands up to 36 inches annually, where you begin to find larger patches of eastern deciduous forest mixed with open prairie. The 20-inch rainfall line and westward is a general agricultural boundary, where good crop production depends on supplemental irrigation. Not to say that modern drought-tolerant crop hybrids cannot survive the drier High Plains, but success is more predictable with irrigation. Keep in mind that as you move northward, cooler overall temperatures and more consistent snow cover mean a shift toward cool-season plants and crops. Conversely, southward, warm-season plants and crops increase in abundance. The transition zone, where warm- and cool-season flora mix, is generally located across southern Colorado and northern Mexico, eastward to the Kansas-Oklahoma border. Again, higher elevations (Denver is at 5280 feet, Wichita, 1300 feet) play a large part in stirring the cool-season/warm-season mix.

Ecology of the Steppe

A varying fire regime is responsible for the immense plant diversity that once survived on the central North American steppe, more so than any other steppe environment in the world. Plants of the vast open grasslands not only evolved to withstand frequent fires, they became dependent upon them: fires cleanse the areas of competing woody vegetation that competes with sun-loving grassland species. Natural fire was more frequent in the tallgrass and mixed-grass prairies to the east and more southerly and warmer prairies to the south, less frequent in the more xeric shortgrass prairies of the West and in northern mixed-grass prairies. Lightning was the frequent igniter of these conflagrations; depending on the time of year, fire promoted grasses over forbs and always discouraged the invasion of the sensitive woody species. In the last 10,000 years, the Native American nomad also learned to use fire to attract wild game by burning prairies in winter and spring; the quick return of tender shoots brought large herds of buffalo and elk within easy range.

The vegetation of the central steppe is dominated by grasses and forbs that die back to the roots each season and resprout in the spring, a mosaic of different types of grasslands based on latitude, elevation, moisture gradients, exposures, soils, and substrates. There are few sharp lines of demarcation; one type of grassland grades into another. More than 2900 species of vascular plants occur throughout this region, many of them recent and adventive, having colonized the region (from the ice-free south, southeast, and southwest especially) after the retreat of the last Ice Age. The central North American steppe is a huge convergence zone of many types of plants that have ranges that extend well beyond the current boundaries. In the northern reaches, what was not covered by ice 10,000 years ago was largely cool temperate forest that retreated northward as the climate warmed and became drier. Resilient grasslands quickly colonized this new territory, but some of these relic species remained in favorable pockets of the plains. Today, well north of the Platte River in Nebraska, scattered pockets of quaking aspen (*Populus tremuloides*) subsist and become more common as you move into South and North Dakota, eventually grading into the Aspen Parkland of the prairie provinces of southern Canada.

The prairies of the region are defined broadly by the height of the grasses

Frequent fires across the central steppe play a key role in plant diversity.

that dominate them. Shortgrass prairie (grasses 15 to 60 cm) in the drier west and southwest is dominated by *Bouteloua gracilis*, *B. dactyloides*, *Sphaeralcea coccinea*, and *Carex duriuscula*. Mixed-grass prairie (grasses 60 to 120 cm) in the somewhat moister central region is dominated by *Elymus smithii*, *Schizachyrium scoparium*, and *Bouteloua curtipendula*. Tallgrass prairie (grasses greater than 120 cm) in the humid eastern stretches of the steppe is dominated by *Andropogon gerardii*, *Panicum virgatum*, and *Sorghastrum nutans*. Many other types of more specialized prairies exist, including sand, sand/sage, northern mixed grass, and southern mixed grass.

A Biogeographic Tour of Central Steppe Flora

Major geologic features of this vast area, intruding above or cutting below the plains that interrupt the gentle eastward-trending downslope, are at times impressive but mostly imperceptible to the casual traveler. To the northwest is the Missouri Plateau, bisected by the Missouri River, a major water highway for the northern portions of the steppe; Lewis and Clark rowed up its length from near St. Louis in 1803, searching for a water passage to the Pacific Coast. The plateau, a glaciated, flat, striated-to-rolling expanse ranging from 1750 to 3300 feet in elevation, receives only 15 to 17 inches of precipitation annually. Cool-season grasses and forbs predominate, chiefly associations of light blue western wheatgrass (*Elymus smithii*) and needle grass (*Stipa* spp.), together with a small complement of warm-season grasses, including little bluestem (*Schizachyrium scoparium*), blue grama (*Bouteloua gracilis*), and sandreed (*Calamovilfa longifolia*) growing on a sandstone and shale substrate interlaced with coal seams. The area is largely dryland farmed and grazed.

Southward, the Black Hills arise in western South Dakota and eastern Wyoming. Largely composed of granite dome that surged upward through the limestone substrate, the Black Hills are a sky island on the plains, harboring Rocky Mountain–like vegetation on the higher elevations. Rising 5500 to 7250 feet and boasting 19 to 24 inches of annual precipitation, they are a stark contrast to the lower prairies. Ponderosa pine (*Pinus ponderosa*), paper birch (*Betula papyrifera*), and quaking aspen (*Populus tremuloides*) carpet moister slopes, grading into cool-season shortgrass prairie at lower elevations. Western white spruce (*Picea* ×*albertiana*), a compact hybrid used in landscaping, is also found in these hills, and the uncommon *Lilium philadelphicum* var. *andinum*, a western version of the wild lily, survives in cool shade along lightly wooded slopes.

Move southeast and you reach the Nebraska Sand Hills, the great filter and funnel of the mighty Ogallala Aquifer, a 450,000-square-kilometer underground reservoir underlying eight states from South Dakota and Wyoming down through Texas and New Mexico. Modern well technology drawing from the aquifer spawned the center-pivot irrigation pattern so distinctive as you fly over today. The Sand Hills themselves occupy 50,000 square kilometers of vegetated dunes formed 5000 to 8000 years ago, as blowing and drifting sands picked up largely from the Goshen Hole settled here. Goshen Hole is a depression along the North Platte, 150 miles long by 50 miles wide, where the river's shifting course constantly exposes drying sand to the prevailing winds. In the Sand Hills, an average of 24 inches of precipitation sinks annually and swiftly into the permeable dunes, feeding numerous local rivers and wetland complexes. The dunes are dominated by sandreed (*Calamovilfa longifolia*), sand bluestem (*Andropogon hallii*), little bluestem (*Schizachyrium scoparium*), hairy grama (*Bouteloua hirsuta*), and needle-and-thread (*Hesperostipa comata*), which grasses support large ranches. The endemic blowout penstemon (*Penstemon haydenii*) survives in sand blowouts caused by pervasive winds; unique in the genus, its stems have the ability to form adventitious roots as they lay over the sand, while a deep tap root anchors the plant. More commonly found in the eastern United States, yellow star-grass (*Hypoxis hirsuta*) grows in moist bottomland hayfields nestled between the dunes

Lilium philadelphicum var. *andinum* grows in moist woodlands in the Black Hills and a few sites along the foothills of the Rockies.

Penstemon haydenii, an endangered endemic, grows in the Nebraska Sand Hills and in the dunes of eastern Wyoming.

The huge petunia-like flowers of *Ipomoea leptophylla* are a common sight in sandy soils throughout the steppe.

along with *Lilium philadelphicum* var. *andinum*. *Iris missouriensis*, extremely common in the wetter montane valleys of the Rockies, reaches its eastern limit here along seeps and creeks. *Ipomoea leptophylla* is a common sight here.

The Platte River provided pioneers a narrow corridor of water and wood in a semi-arid grassland. The Mormon and Oregon trails followed it, as did the transcontinental railroad in the late 1860s. Its two major branches, the South Platte and the North Platte, converge in southwest Nebraska at North Platte after taking wide loops south and north through Colorado and Wyoming, respectively, then continue across the southern tier of the state, emptying into the Missouri just south of Omaha. The shallow, braided stream provided lush, reliable grazing and firewood for wagon trains, from willows and plains cottonwood (*Populus deltoides* ssp. *monilifera*). Myriad wetlands for major migrating flocks, from sandhill cranes to Canada and snow geese, make this critical wildlife habitat. Grasses (*Elymus smithii*, *Andropogon gerardii*, *Stipa* spp.) dominate the uplands along the central portion of the river.

Arising in northeastern Colorado and looping up into southwest Nebraska, the Republican River courses through the low, red eroded hills near Red Cloud, briefly the home of author Willa Cather, before turning southeast to join the Smoky Hill River in Kansas, forming the Kansas River. This region of Nebraska is home to mixed-grass prairie, which deeply influenced Cather's writing. Both resinous skullcap (*Scutellaria resinosa*) and golden prairie clover (*Dalea aurea*) are mixed-grass denizens, growing on well-drained slopes and ridges. The mixed-grass prairie is where the naturally dwarf *Baptisia australis* var. *minor* appears in

Sandhill cranes migrate with the seasons in the central portions of the steppe.

the west and continues eastward through much of the Flint Hills. This is the northern limit of *Scutellaria*'s range; the genus is more common southward into Texas and west into Arizona. Golden prairie clover ranges from western South Dakota to the Gulf Plains of Texas, and a disjunct population occurs in Arizona. Highly palatable to cattle, *Dalea aurea* is only recently restricted to localized, lightly grazed prairies and shrublands.

Fremont's evening primrose (*Oenothera macrocarpa* ssp. *fremontii*) grows around Red Cloud as well, but the center of its diversity is just to the south, the Smoky Hills of north-central Kansas. Some of the most dramatic and picturesque scenery in Kansas, the rolling and rocky hills are dominated by mixed-grass prairies. Sideoats grama and little bluestem share well-drained sites with Fremont's evening primrose, which can create dazzling displays of yellow along road cuts and rocky ridges on early June mornings. A western form of the Missouri evening primrose, Fremont's is more compact and drought tolerant with very linear, blue-green leaves. Its range extends south and west to eventually intergrade

Fremont's evening primrose grows commonly in limestone gravel and rocky areas in the Smoky Hills of north-central Kansas.

with *O. macrocarpa* ssp. *oklahomensis* in the Red Hills of south-central Kansas and *O. macrocarpa* ssp. *incana* to the southwest. The endemic Fremont's clematis (*Clematis fremontii*), a non-vining member of the genus with frosted blue-violet cupped flowers, occurs in only six counties straddling the Nebraska-Kansas border; in this area, Fremont's clematis is one of the most common plants in the prairie, and where it occurs, it can be considered a weed. I have seen solid masses growing so densely, it literally crowds out all other vegetation along ridges in Lincoln County, Kansas. Another isolated population of this plant occurs east in Missouri on some of the limestone glades, where the rocky soils prevent growth of eastern deciduous forest; this eastern variety varies morphologically by having much more acute, linear leaves.

To the east and south, Dakota sandstone prairie stretches from southeast Nebraska down into central Kansas. Birds-foot violet (*Viola pedata*) makes showy purple patches in early spring contrasting against still-dormant grasses. The Maxwell Game Preserve in east-central Kansas is as far west as it ranges; it is more at home in the moist upper Midwest glades and open woodlands.

The Arkansas River arises just west across the Mosquito Range from the South Platte River in the central Colorado Rockies, and another critical highway for settlement, the Santa Fe Trail, followed it up through Colorado. Accounts from as early as the 1850s document the overgrazing and destruction of the cottonwood galleries and the rich bottomland prairies along the river by hordes of

Fremont's clematis growing along a hillside in north-central Kansas.

Spring wildflowers in full bloom on sandy lowlands of the Arkansas River Valley, just west of Dodge City, Kansas.

settlers, miners, traders, and the horse-hoarding indigenous tribes, who wintered their herds in the sheltered river valley and fed their horses bark from the cotton-woods. The famous Big Timbers of the Arkansas River, which harbored dense stands of old growth cottonwood (*Populus deltoides*) along its banks, was largely destroyed by 1853 (West 1998).

Nowadays, the Arkansas ceases to flow for many miles past the Colorado border except for the increasingly rare wet years in the Arkansas River Valley. Through Garden City and Dodge City, Kansas, it is largely a sandy wash—a victim of intense irrigation demands upriver in Colorado and western Kansas, coupled with a decrease in the level of the Ogallala Aquifer from irrigation pumping. Sand/sage prairies are abundant along its flanks through Colorado and Kansas, formed by the wetting and drying of the river beds and persistent winds scooping out the sand deposits. In certain stretches of eastern Colorado and western Kansas, spring wildflower blooms, dominated by short-lived perennials and true perennials, can be spectacular, depending on available moisture: the silver foliage of *Artemisia filifolia*, the orange-red flowers of *Sphaeralcea coccinea*, multi-colored *Gaillardia pulchella*, and the creamy blooms of the ubiquitous *Yucca glauca* create large colorful mosaics across the low, sandy prairies.

The eastern *Hypoxis hirsuta* found in the Nebraska Sand Hills also occurs in moist pockets of Sand Hills State Park in central Kansas. These dunes, derived from the Arkansas River flowing just to the southwest, feature numerous ponds and small creeks supporting lush wetland vegetation and extensive forest galleries. The biennial *Oenothera heterophylla* var. *rhombipetala*, its tall dense spikes loaded with clear yellow flowers, grows here and in other sandy prairies of the

The heavy-blooming rough dogwood is common along ditches and in draws across much of the eastern and central portions of the steppe.

RIGHT Purple and white prairie clovers are widespread across the steppe.

central steppe, along with the widely cultivated lavender *Penstemon grandiflorus*. *Dalea villosa*, with its densely tomentose, silvery foliage and rose-purple flowers, is frequent on the upland dunes.

The fine sand blown over impermeable clay substrate traps moisture as it flows down, creating a perched water table. Sand bluestem, sand dropseed, and little bluestem dominate the uplands; prairie cordgrass (*Spartina pectinata*) and numerous sedges and rushes dominate lowlands. Extensive stands of the extremely floriferous rough dogwood (*Cornus drummondii*) and Chickasaw plum (*Prunus angustifolia*), the sweetest of all the wild plums in my opinion, dominate large areas along the lower edges of the dunes. Extensive tree galleries along the creeks and draws feature *Catalpa speciosa*, *Celtis occidentalis*, *Fraxinus pennsylvanica*, and massive *Populus deltoides*.

Black walnut (*Juglans nigra*) thrives in these moister sites. It is also common along the Platte River, the Arkansas, and the Cimarron to the south, where it hugs the waterways, allowing it to penetrate farther into the drier west than it would normally be able to grow. The venerable bur oak follows this same pattern to the north. A logger from southeast Kansas told me about large stands of bur oaks along the Saline River lowlands just south of Hays, on the 100th meridian. *Quercus macrocarpa* also occurs well into Wyoming just east of Devils Tower.

An eroded complex of uplands, deeply cut by canyons, feeds streams near the headwaters of the Salt Fork of the Cimarron River. Clark State Lake of southwest Kansas rests in a deep canyon eroded down into the Ogallala limestone formation here. The upper edge of bluffs has weathered into limy gravel. Hoary evening primrose (*Oenothera macrocarpa* ssp. *incana*) thrives in this narrow band of

Amorpha nana growing in a garden with *Phlox nana*.

limestone grit, tracing the edges of the canyon. Scattered around are a contingent of southwestern natives, including lavender-leaf sundrops (*Calylophus lavandulifolius*) with its crinkled, translucent yellow flowers and the incredibly beautiful Irish eyes (*Berlandiera texana*). *Dalea aurea* reappears here, along with a more recumbent form of purple prairie clover (*D. purpurea*) on the better sand/silt soils. Both *D. purpurea* and *D. candida* are widespread throughout most prairie types over much of the steppe; tall and upright in the east, they get shorter and more sprawling in the drier regions of the central steppe, regaining their stature only as they near the Front Range.

Deep in the canyon, restricted to a band of shale near the bottom, the uncommon *Amorpha nana* grows; this small remnant population is 300 miles from where it occurs on the mountain slopes above Boulder, Colorado, at around 7000 feet in elevation, and even farther from its upper Midwest populations. Along the ravines of the canyon, *Juglans microcarpa* reaches 10 to 15 feet tall. Just to the east, *Sapindus drummondii* grows along streams, and to the west, individual tortured specimens of western hackberry (*Celtis laevigata* var. *reticulata*)

Eustoma exaltatum grows along the edges of ephemeral ponds in the southwestern portions of the steppe.

begin to appear along rocky ridges, where lack of vegetation protects them from frequent fires.

Such deep canyons harbor amazing wildlife. I grew up in the mule deer country of mountainous south-central Colorado, but the largest bucks I have ever seen ran out ahead of me from Clark State Lake one fine June morning. Their antlers were massive, probably owing to the mineralization in the sand/silt soils, and their bodies rippled with muscles and fat, due to the rich shrub browse in the canyons and frequent visits to nearby milo fields on the rolling uplands. These southwestern mixed-grass prairies of Kansas are some of the most stunningly beautiful I have ever viewed if the rainfall has been sufficient. Rich waves of sideoats grama, little bluestem, and blue grama traipse over gently rolling hills that suddenly drop into near vertical cut canyons harboring rich varieties of trees, shrubs, and dazzling wildflowers. Many researchers who have bothered to look at mixed-grass prairies claim it is more floristically diverse than either tallgrass or shortgrass prairies. True, it is a transition zone between these two major types, but it also has a distinct set of plants that occur only within its confines.

To the west another 60 miles, where the Cimarron loops up through Kansas, low rolling sandy hills flank the river just north and east of Elkhart, Kansas. Tree cholla (*Cylindropuntia imbricata*) reaches upward of 10 to 12 feet, the tallest I have seen it anywhere in its range. It looks strangely out of place in the rolling plains devoid of trees or any plant that rises above 20 inches tall. Amazingly, much of the surrounding sandy prairie was once plowed and farmed, but

Tree cholla is widespread in the southern portions of the steppe.

the 1930s Dust Bowl turned it into a breeding ground for massive dust storms. Tree cholla continues west, up to the foothills south of Colorado Springs, then retreats to Arizona and south to Mexico. Strangely, south of the Cimarron River in the Oklahoma Panhandle, it is extremely infrequent and tends to be squat and bushy, rising only to about a meter, similar to what it looks like in south-central Colorado, except it is more densely branched with a much greater circumference. The yellow flowers and the short, spineless fruits provide a food source to many birds and deer. Today, largely due to government revegetation efforts on abandoned farmland in the late 1930s and '40s, Cimarron National Grasslands occupies the flanks of the river and provides rich habitat for everything from antelope to mule deer and white-tail deer. Surprisingly in this arid landscape, scattered around the edges of ephemeral ponds created by clay layers underlying the sand, one can view large swaths of *Eustoma exaltatum* in full bloom midsummer.

The Canadian River cuts a deep swath across the northern border of the Caprock Escarpment on the northern edge of the Llano Estacado; for much of its course across eastern New Mexico and Texas, uplands along the river are dominated by buffalograss (*Bouteloua dactyloides*), galleta grass (*Pleuraphis jamesii*), and burrograss (*Scleropogon brevifolius*), gradually transitioning into buffalograss

Caprock limestone protruding through the grasslands of New Mexico.

and honey mesquite in Oklahoma. Fragrant mimosa (*Mimosa borealis*) is common in rocky areas. Berlandier's sundrops (*Calylophus berlandieri*) and lavender-leaf sundrops are found here, as well as scattered *Juniperus monosperma* and more *Celtis laevigata* var. *reticulata* (McGrath 2010).

One of the world's flattest regions, the Llano Estacado slopes less than 10 feet per mile from its high edge in New Mexico, at 5000 feet, to the Permian Plains of Texas, at 3000 feet. Rising nearly 300 feet above the Caprock Escarpment, it is composed of hard limestone, a further testament to ancient sea beds. The soft, uneven soils originally were dominated by warm-season buffalograss/blue grama shortgrass prairie, but since the 1950s the deeper soils were subject to wholesale pumping for water-intensive crops like cotton; this dropped the water table significantly, jeopardizing the economy of the region.

The major geologic feature of the eastern border of the steppe is the Flint Hills, arising in northeast Kansas, just south of the Nebraska border, and trailing down into Oklahoma's Osage Hills. The region hosts the last great tract of intact tallgrass prairie in North America. Chert nodules laid down with limestone during the Permian period made the substrate less prone to erosion, and the resulting soils are gravel/chert mixture. At this eastern, more humid edge of the steppe, annual rainfall ranges 32 to 36 inches per year, which supports dense stands of big bluestem, Indiangrass (*Sorghastrum nutans*), and switchgrass (*Panicum virgatum*), along with such classic tallgrass prairie forbs as butterfly milkweed (*Asclepias tuberosa*), pale coneflower (*Echinacea pallida*), and prairie sage (*Salvia azurea*). *Nemastylis geminiflora* (prairie iris) grows here in moist low-lying areas,

LEFT Iconic and beautiful Indiangrass is common in the Flint Hills and more humid prairies of the eastern edge of the steppe.

BELOW The picturesque Flint Hills border the eastern edge of the steppe in Kansas.

as well as in drier hayfields and pastures; prairie plantsman Claude Barr (1986) called it a glorified sisyrinchium.

Here on the eastern frontier of the steppe, eastern deciduous forest fills the draws and rivers valleys, dominated by oaks (bur, chinquapin, black, and post) but with strong contingents of black walnut, cottonwood, sycamore, redbud, coffee tree, American elm, and green ash. With every ridge crossed to the east, the forest becomes denser.

Woody Plants of the Steppe

Woody plants are the exception in much of the central steppe. The majority of woody trees and shrubs are restricted to islands—bends in waterways, canyons, mesa tops—where they are protected from the frequent fires that historically swept through the region. Leadplant (*Amorpha canescens*) can survive in open grasslands by being burned back to the crown and resprouting from its deep root system; once widespread through short-, mixed-, and tallgrass prairie, when you see it today, it is almost always indicative of diverse plant populations, as it has been grazed out of most of the plains by cattle and sheep.

The suppression of fires across the plains has led to an increase in woody vegetation in many places. One example of a taxon whose range has expanded due to absence of fire is the ubiquitous *Juniperus virginiana* (eastern red cedar), which occurs sporadically throughout the eastern half of the steppe and is in fact the only conifer indigenous to Kansas. An extensive forest of eastern red cedar covers north-central Oklahoma and the Red Hills of south-central Kansas. Other notable woody populations occur in eastern Montana and Wyoming, where *Rhus trilobata* and *Artemisia tridentata* ssp. *wyomingensis* dominate ranges with adequate moisture and soil conditions.

The same patchiness holds true for localized dominance of honey mesquite (*Prosopis glandulosa*) in western Oklahoma and the Texas Panhandle and Permian Plains. In the panhandle, honey mesquite dominates the draws and canyons near the Canadian River, but up on the exposed plains it disappears. Honey mesquite increases in dominance under intense grazing, forming dense stands that crowd out grasses, making it a pest species for range managers (although the pods are nutritious browse to livestock and wildlife alike).

Sand shinnery, a dense shrubland dominated by rhizomatous *Quercus havardii*, occupies rangeland along the sandy flanks of the Canadian River starting in central New Mexico and spreading into western Oklahoma. I have attempted to walk through sand shinnery in Gray County, Texas, and it is virtually impenetrable, prompting many ranchers and landowners to bulldoze the shrubs to make way for range grasses. Sand shinnery's density brings new meaning to the term "shrub steppe"!

RIGHT Dense stands of *Quercus havardii* dominate specific areas of the Texas Panhandle and eastward into western Oklahoma.

BELOW Mesquite growing along the Canadian River in the Texas Panhandle.

In eastern Colorado, headwaters of the Arikaree River support ponderosa pine and Rocky Mountain juniper.

Just to the southwest, remnant populations of *Juniperus scopulorum* (Rocky Mountain juniper) intergraded with *J. monosperma* and *J. deppeana* occur in heavier soil and moisture gradients in the Palo Duro Canyon of Texas. Rocky Mountain juniper also spills out from the mountains in eastern Montana and western North Dakota, as far east as the Badlands of South Dakota.

In the central steppe, the Niobrara River cuts a 400-foot-deep canyon through north-central Nebraska, exposing many different soil types while providing an array of microhabitats that harbor several species not generally found on the steppe. On the cool south sides of the canyon, paper birch (*Betula papyrifera*) and a naturally occurring hybrid aspen (*Populus ×smithii*) create an overstory; as you move lower, hophornbean (*Ostrya virginiana*), *Tilia americana*, and *Corylus americana* form an extant eastern deciduous forest well west of the 100th meridian! On the lower sub-irrigated river terraces, eastern tallgrass prairie and bur oak dominate. On the north side of the river, where the sun intensity is stronger, western coniferous woodland dominated by ponderosa pine grades into northern mixed-grass prairie. This confined area supports a fascinating mixture of ecotones.

A few peninsulas of treed ridges extend out into the plains from the Rocky Mountains. Palmer Ridge in central Colorado is dominated by ponderosa pines and Gambel oak. Ponderosa pine hugs the ridge well out into the eastern plains of Colorado, disappears for about 20 miles as you move eastward, then recurs along with Rocky Mountain juniper for another 15 miles or so along the headwaters of the Arikaree River just north and west of Limon, Colorado. The deep

Single-seed juniper and tree cholla dominate the dry tablelands of southern Colorado and parts of northern New Mexico.

Limber pine finds a safe harbor on the shaded slopes of Pawnee Buttes.

canyons and cool north-facing slopes of this area are a surprising oasis on the high plains, a haven for conifers well beyond where you would expect to find them.

Pine Ridge, another treed peninsula, continues out from the Black Hills to hug the border of Nebraska and South Dakota. Palmer Ridge acts as a dividing line between southern desert species and northern species. To the south, piñon pine and single-seed juniper are dominant, but both disappear to the north. Tree cholla is common in the foothills of New Mexico and southern Colorado, but it too becomes very scarce north of the ridge.

North of Palmer Ridge, cool-season grasses are more common as is Rocky Mountain juniper. *Juniperus horizontalis* and *J. communis* follow suit, growing out into the Northern Plains. Pawnee Buttes in northeast Colorado has an isolated population of *Pinus flexilis* (limber pine) on its cooler shaded slopes; otherwise, apart from a few spots along the Raton Basin, this taxon is found only at higher elevations southward.

WOODY PLANTS OF THE STEPPE

Colorado's Fishers Peak—at 9600 feet, the highest point on the Raton Basin's western edge.

The Raton Mesas is an area of lava-capped plateaus and buttes dominated by piñon pine (*Pinus edulis*), single-seed juniper (*Juniperus monosperma*), and to a lesser extent *Quercus gambelii* and *Q. undulata*, extending as far east as the tip of the Oklahoma Panhandle on the Black Mesa in Cimarron County. The junipers become scattered and gradually disappear just north of Boise City, Oklahoma, and small scattered stands can be seen traveling north into Baca County, Colorado. A small population of ponderosa pines exists on Black Mesa, which is as far east as they range in the Southern Plains. *Yucca baccata* and *Prosopis* do not grow north of the Raton Basin, which acts as a barrier to more desert species.

Montane Steppe

The steppe does not end on the eastern flank of the Front Range. Surprisingly, the arid grasslands can be found at higher elevations along the Rockies, where localized climatic conditions brought on by high peaks to the west and warmer southern exposures create dazzling prairies above 10,000 feet in elevation. The Wet Mountains of southern Colorado are covered with dry grasslands along the drier west slopes with south-facing exposures. Below, to the west, is the Wet Mountain Valley, a thriving cattle and hay ranching area, which depends on snowmelt runoff from both the Wets and the Sangre de Cristo to irrigate rich wet meadows.

Once out of the valley, the dry grasslands are dominated by rabbitbrush

Colorado's Wet Mountains support dry grasslands interspersed with bristlecone pine up to 10,000 feet in elevation.

(*Ericameria nauseosa*), sulphur flower (*Eriogonum umbellatum*), and several paintbrush species (*Castilleja linariifolia, C. integra,* and *C. miniata*), among others. As you move upward in the Wet Mountains, the near-treeless valley gives way to denser stands of ponderosa pine, piñon pine, and a host of shrubs, including fourwing saltbush (*Atriplex canescens*) and alderleaf mountain mahogany (*Cercocarpus montanus*). Above 9000 feet, you encounter groves of *Populus tremuloides* with *Aquilegia caerulea* and other denizens of the high grasslands; even at these higher elevations on the drier exposed slopes, one still encounters *Bouteloua gracilis* and other familiar steppe species. Blue grama is pushed to mature in a growing season that can be less than 40 days because it is technically a warm-season grass. Other steppe species found at the highest elevations include *Oxytropis lambertii, O. sericea,* and several penstemons. Small creeks

Yellow lady's slipper orchid growing in wet meadows along the valley floor of Colorado's Wet Mountain Valley.

in the area support wetland vegetation, including *Dodecatheon pulchellum* and *Dasiphora fruticosa*. The latter is another circumboreal species that occurs in moister areas of high-elevation steppe throughout the northern hemisphere; I recently saw it growing abundantly just northeast of Ulaanbaatar, Mongolia.

Along with an abundant numbers of elk and mule deer at these elevations, pronghorn antelope, the speedster of the prairies, inhabit this high steppe grassland during the short summers to give birth to new kids and graze on the lush, mostly cool-season grasses.

Plant-People Connection

For centuries, settlers and explorers of the central North American steppe viewed this open expanse as empty, dry, and featureless—a place to be endured while crossing to the furs, gold, and timber of the West. The perception of emptiness continues today—seemingly endless, flat expanses of farms and ranches bisected by arrow-straight roads, occasionally interrupted by a small town—but this emptiness is an illusion brought on by a challenging topography. The steppe holds a remarkable flora of hardy, widely adaptable, drought-tolerant plants that in recent decades has drawn increased scrutiny by the horticultural community. The demands for water-efficient plants, fueled by expanding cities in the arid West and Southwest, have brought plants from steppe regions around the world into the built landscape. James Stubbendiek captured the enigma of the plains. "It seems to be a constant contradiction of itself. It is delicate, yet resilient; it appears to be simple, but closer inspection indicates that it is extremely complex; it may appear monotonous, but it is diverse and ever-changing throughout the seasons" (Cushman and Jones 1988).

The plains, with few prominent features and rivers that dry to a trickle in late summer, were and remain in sharp contrast to the dramatic landscapes and heavily forested reaches west and east of them. In 1823, geographer Edwin James wrote of the region he'd labeled "The Great American Desert": "[I]t is almost wholly unfit for cultivation, and of course, uninhabitable by a people depending upon agriculture for their subsistence." James had already been conditioned to think this way: when the region was obtained by the United States as part of the Louisiana Purchase in 1803, President Jefferson wrote of its "immense and trackless deserts." In 1810, Zebulon Pike, for whom Pikes Peak is named, ventured a prediction: "These vast plains of the western hemisphere, may become in time equally celebrated as the sandy deserts of Africa." And his map of the region included this annotation: "not a stick of timber" (Meinig 1993). In *Roughing It* (1872), Mark Twain summed up the South Platte River in eastern Colorado: "We came to the shallow, yellow, muddy South Platte, with its low banks and its scattering of flat sand-bars and pigmy islands—a melancholy stream straggling through the centre of the enormous flat plain, and only saved from being impossible to find with the naked eye by its sentinel rank of scattering trees standing on either bank. The Platte was 'up,' they said—which made me wish I could see it when it was down, if it could look any sicker and sorrier."

In the latter half of the 19th century, burgeoning populations of land-hungry immigrants, relentless promotion by railroads, and near complete suppression of Native American tribes moved homesteaders out into the plains, where they began to adapt to the unfamiliar, treeless environment. Pioneers emerging from the western edge of the eastern forest recount the spectacle of a sea of grass, a feeling Willa Cather (1918) captures: "As I looked about me I felt that the grass was the country, as the water is the sea. The red of the grass made all the great prairie the colour of wine-stains, or of certain seaweeds when they are first washed up. And there was so much motion in it; the whole country seemed, somehow, to be running." What these pioneers discovered was that commodity crops like wheat, corn, milo, and alfalfa thrived in the incredibly fertile soils of these open spaces. The Great American Desert slowly transformed into an agricultural breadbasket.

The scale of that transformation has been immense, and the effect on the native grasslands has been traumatic to the ecosystem. The grasslands and scattered woodlands that remain are mere patches of the biodiversity that once thrived here, but also a testament to the resilience of a flora adapted to extremes. The populations of large fauna that depended on the grasslands have been severely reduced, confined to game preserves and tightly managed private herds. The great Central North America steppe has been almost entirely domesticated.

A herd of buffalo grazes on the Maxwell Wildlife Refuge in central Kansas.

Animal Influence

It is remarkable how many early accounts of this region make no mention of the massive herds of elk and buffalo that roamed the "desert." An estimated 30 to 50 million buffalo once ranged from Mexico to Canada, and the herds supported populations of nomadic people who followed their migrations. By the late 19th century, only about a thousand free-ranging buffalo remained. The buffalo herds were a major ecological influence on the grasslands. Buffalo (*Bison bison*) prefer to graze grasses and largely leave forbs alone. As the huge herds seasonally moved north and south across the steppe, they tended to intensively graze the grasses out of an area and then move on, leaving an aerated and fertilized ground loaded with untouched forbs to reseed the area. Prairie fires, started by lightning or set by nomadic Indian tribes, cleared the prairies of overlying dead vegetation matter, promoting faster nutrient cycling and quicker green-up in spring. In the early 20th century, an estimated 5 billion black-tailed prairie dogs inhabited the plains. Prairie dogs are a keystone species, providing food for several predators

Coyote pups at play near
the safety of the den.

Elk were once abundant on the grassland steppes from the Front Range
of the Rocky Mountains to the tallgrass prairies on the eastern edge.

and habitat for other ground-dwellers. Farming and the perception that prairie dogs destroy the grasslands led to relentless efforts to limit their range. The loss of the prairie dog played into the loss of the predators that relied on them as an abundant food source.

Elk (*Cervus elaphus*), now mostly thought of as mountain-dwellers, were once common on the steppe as well. Like buffalo, they are grass grazers, but like deer, they also browse a wide variety of shrubs and forbs. The great elk herds were hunted to near extinction on the plains in the 19th century, driven westward into the back country of the Rocky Mountains and Great Basin, where they still exist in large numbers. A few remain on grassland reserves on the plains. The elimination of the buffalo and elk herds, coupled with the suppression of fires and the introduction of cattle (non-selective grazers, especially when confined by fences), has greatly decreased the forb diversity of the prairies and allowed for greater grass and woody plant dominance.

Other animals we no longer associate with the steppe, although they were once common there, are gray wolves (*Canis lupus*) and grizzly bears (*Ursus arctos*). The abundance of large herds of animals, both large and small, made readily available food sources for these voracious predators. Human encroachment, the steady replacement of wild herds with domesticated cattle, sheep, and horses, and the constant pressure by humans to eliminate these predators, pushed both the wolf and grizzly into the less settled Pacific Northwest, where there were sufficient

Plant-rich prairies are a rare sight in the modern steppe but the look lives on in prairie-style gardens.

wildlands and remaining wild herds to support them. The coyote (*Canis latrans*), the prairie wolf to Lewis and Clark, has opportunistically adapted to the changing landscape and food sources of the steppe, despite the best efforts of settlers to eliminate it. This amazingly adaptable species maintains a solid foothold in rural areas across the plains and thrives in suburban and even in urban habitats.

The prairies of the central North American steppe once boasted areas with as many as 300 different plant species per acre. Today, with the almost complete suppression of the ecological forces that shaped them, diverse grasslands are extremely rare and generally confined to a few protected acres scattered across the steppe. In the over ten years I spent traveling and exploring the central steppe, I can honestly say the best natural prairies I saw may have had at most 150 species per acre, and that was only in tiny patches that had somehow evaded severe grazing or plowing due to a quirk in topography or an anomaly in fencing. Case in point: near the small hamlet of Cambridge, Kansas, in the southern Flint

Tallgrass Prairie National Preserve contains one of the last large tallgrass prairie tracts in the central North American steppe.

Hills, a tiny patch of tallgrass prairie survives along a small swale with deep soil, nestled into the west face of a rocky ridge. I never really could imagine what true tallgrass prairie looked like until I discovered this remnant over 20 years ago. Less than two acres in size, this patch contained almost every prairie plant I associated with tallgrass prairies and several I had never seen before! Purple and white *Camassia scilloides* tucked in next to *Lithospermum onosmodium*, rusty red *Asclepias tuberosa* with hot pink *Mimosa quadrivalvis* poking up through it; the Tallgrass Prairie National Preserve near Strong City, Kansas, has a good base of prairie species, but nothing I have seen on that property approaches the diversity I observed on this mere two acres. It will take years and years of plant restoration and reintroduction of the natural ecological forces to even begin to approach a more natural state of the original prairie. "What a thousand acres of compass plant looked like when they tickled the bellies of the buffalo is a question never again to be answered, and perhaps not even asked" (Leopold 1949).

ANIMAL INFLUENCE

Asclepias tuberosa, the well-named butterfly milkweed, is the norm in sand prairies.

Asclepias viridis, one of the many *Asclepias* species that deserve wider cultivation in gardens.

CENTRAL NORTH AMERICAN STEPPE

APOCYNACEAE

The flowers of Apocynaceae are not only subtly beautiful but extremely complex. Pollen is packaged in a structure resembling saddlebags. Pollination occurs between plants when insects such as bees hook the structure on their leg spurs and fly it to adjacent plants and properly place it in a specialized groove on another flower. Specialized, but it obviously works, as witnessed by thousands of floating milkweed seeds wafting in the breeze across the steppe every fall!

Many *Asclepias* species deserve wider cultivation in gardens. *Asclepias viridis* (green antelope horn) is up to 24 inches tall; the umbel inflorescence can reach 5 inches across and boast upward of 18 individual flowers. Flowers have green sepals, green petals, and red to purple hoods. The fruits are always an attraction, forming a pod 3 to 6 inches long stuffed with abundant seeds with long white hairs—individual sails for seed dispersal. Prefers sandy or rocky prairies, often on calcareous soils. *Asclepias asperula* (spider antelope horn) is similar but with leaves more narrowly lanceolate. Besides the classic *A. tuberosa*, other noteworthy milkweeds of the steppe include the tall, linear-leaved *A. engelmanniana* and *A. verticillata* (whorled milkweed), with grass-like leaves arranged in whorls about the stem.

ASTERACEAE

Berlandiera texana (Irish eyes), the lesser-known cousin of chocolate flower (*B. lyrata*), is much more attractive and every bit as rugged, hailing from the sandy windswept prairies of the southwestern steppe. Reaching about 26 inches tall, hosting cordate, clasping leaves and yellow to orange-yellow widely spaced ray flowers May through July, this is a gorgeous species that merits a spot in the garden.

Pale coneflower is the most delicate of the coneflowers while blooming in early summer; later, its dried seedheads provide excellent feed for overwintering birds.

Tall, dainty *Echinacea pallida* (pale coneflower) deserves wider cultivation for its extremely long, linear ray florets. In late June, the delicate composite heads appear like badminton shuttlecocks in the xeric garden. So long and dainty are the ray flowers, it is often called spider flower. Reaching as tall as 28 inches, multiple flower stalks with minimal leaves are held aloft over a basal rosette of linear leaves. It used to intermingle extensively in the western edge of the Flint Hills with *E. angustifolia*, a more western and diminutive species, and the aberrant *E. atrorubens*, with the extremely short pale pink ray florets. In the mid-1990s the popularity of echinacea as a cure-all herb prompted most anyone with a shovel to dig them up for profit. They called it snakeroot because if you bit into a freshly dug echinacea root, a little explosion went off in your head, as if you had been bitten by a rattlesnake. I traversed these same areas in 2011, and, sadly, populations had not recovered from the depredation.

Populations of *Gaillardia aristata* growing in Colorado's Wet Mountain Valley are known for their large flowers, robust habit, and strongly perennial constitution.

Echinacea angustifolia (blacksamson coneflower) is shorter, about 16 inches tall, than *E. pallida* but often overlaps its native range, in some areas growing side by side with pale coneflower along the far eastern edge of the central steppe. Tough and rugged, it is more drought tolerant than other coneflowers, and it displays shorter, broader ray florets than *E. pallida*, usually of a deep pink to pinkish red. Tough, glaucous green foliage arising from a stout taproot makes *E. angustifolia* very long-lived as long as soil drainage is better than average.

Gaillardia is well represented in the steppe by *G. aestivalis* and *G. pulchella*, which tend to be annual or short-lived perennials; both are already widely grown, especially in the wildflower seed trade. One of the real gems is *G. aristata*, common in the Northern Plains but tending to hug the Front Range as it comes down into Colorado. *Gaillardia aristata* 'Amber Wheels' is a selection by Jelitto Perennial Seeds based on seed I collected in Colorado's Wet Mountain Valley. A long-lived perennial with large composite flowers, up to 24 inches tall, it thrives in dry soils but also tolerates less-than-great drainage.

Let's look at those fascinating sunflowers, of which the central steppe has an amazing array. With a precise symmetry of ray flowers, *Helianthus tuberosus*

Symphyotrichum fendleri 'My Antonia'.

LEFT Foliage of willowleaf sunflower.

(Jerusalem artichoke) is a stunning, clear yellow sunflower that grows in low seeps and along streams throughout much of the steppe. It has long been in cultivation for its tasty tuberous roots (a.k.a. duck potatoes) but seldom is grown ornamentally, as it tends to be annoyingly invasive. Speaking of thugs, *H. maximiliani* is another robust, rhizomatous sunflower that is widely cultivated for some reason. *Helianthus mollis* (ashy sunflower) is a more compact and less aggressive sunflower more deserving of cultivation for its gray-green, densely tomentose, heart-shaped leaves. My favorite of all these has to be *H. salicifolius* (willowleaf sunflower), with its smaller, abundant composite blooms and delicate, lacy, linear leaves. Mildly rhizomatous, it will hold its own against more aggressive grasses and perennials, owing to experience gained in its native haunts on the limestone prairies of the Flint Hills. Its presence in the garden is magical—pillars of fine-leaved foliage and then a welcome, late summer bloom. Sunflowers are present all the way west across the steppe and north and south, too, with the annuals *H. annuus* and *H. petiolaris* filling any disturbed soil across the plains and into the foothills. *Helianthus pumilus* is an interesting smaller perennial species that hugs the Rocky Mountains, especially in Montana, Wyoming, and Colorado.

Native to the drier portions of the southwestern steppe, *Symphyotrichum fendleri* (Fendler's aster), a low-growing (to 12 inches) subshrub with bluish purple flowers, is perfect for xeric gardens. Tough and compact, it blooms late July well into October, depending on growing conditions. The heavy-blooming selection 'My Antonia', out of Nebraska, deserves wider cultivation.

Tradescantia tharpii creates delightful mounds of color in spring.

COMMELINACEAE

The wispy monocots in the spiderwort family are most noticeably represented by *Tradescantia* in the central steppe. Most of these are common in cultivation, with their grass-like leaves and three-petaled flowers, usually in blue to blue-violet. Ohio spiderwort (*T. ohiensis*) ranges from the east well into the western portions of the central steppe; *T. occidentalis* ranges more in the west, from Canada to Texas, especially in dry sandy soils. The moisture-loving *T. bracteata*, which is tough to tell from the other species, skirts round the edges of the steppe from Montana, across the Dakotas, arcing around to the east back to Illinois, and eventually southwest back into Texas. The most interesting of the lot to me is *T. tharpii* (dwarf spiderwort), which is largely restricted to sandy prairies and rocky outcrops through much of the central portions of the steppe; it is a spring bloomer (late March through April), boasting the widest variation of flower color in the spiderworts, from true blue to violet, pink, and rose-pink. The grassy tufts with a hairy covering, mostly less than 6 inches in height, explode with color in early spring, making them easy to spot scattered across the prairies. Cultivated in well-drained, very slightly rich soils, this plant comes into full glory, creating foot-wide mounds of color!

FABACEAE

Baptisia australis var. *minor* (dwarf blue false indigo) is a delicate, diminutive version of blue false indigo found only in southeastern Nebraska south through the eastern two-thirds of Kansas, extreme western Missouri, Oklahoma, and the eastern regions of the Texas Panhandle. This tough and adaptable legume has light blue foliage followed by spikes of light to dark blue flowers in June. The plant is rarely over 18 inches tall, but the spike of flowers may grow to 26 inches during good growing years on very good soils. And the attraction does not stop there: after blooming, light green pods form, eventually turning coal-black, extending the ornamental qualities into fall. The pods are often used in dried arrangements. Eventually, as the plant senesces, the entire support stem breaks off, and you will see plants tumbling across the prairies in late fall breezes.

Daleas, especially white and purple prairie clovers, have been popular in cultivation for decades. Golden prairie clover (*Dalea aurea*) has never been widely cultivated, attributable in part to its short-lived perennial nature. It boasts dense arrays of clear yellow flowers, which bloom progressively from bottom to top over a three-week period. Next, densely hairy seeds form a very ornamental broad cone (hence another common name, silk-top dalea). The foliage too is very attractive, a silvery green thanks to a densely tomentose covering. This species is palatable to livestock, which contributes to its being surprisingly uncommon over its native range; but where it does occur, its mid-June blossoms are a welcome

Dwarf blue false indigo is a great compact addition to the garden for both flowers and foliage color.

Dalea villosa.

sight, appearing well before the sunflowers put on their prairie show. Persisting for two to four years, like many of the gaillardias, golden prairie clover could be a welcome addition to a dry garden. Just remember to add new plants every few years, to keep a good population of this prairie denizen.

Reaching about 18 inches tall, *Dalea multiflora* (round-head dalea) thrives in the heat of the summer on well-drained rocky and gravelly sites. During the hottest parts of July and early August, this beautiful dalea sends up eight to 12 stems in a refined vase-like habit topped with a profusion of clear white flowers arranged in round pom-pom clusters. It makes a great addition to the dry rock garden and easily thrives on sun-baked south slopes, being especially tenacious because of its woody base and deep root system. An added benefit: the dainty foliage exudes a very pungent, sweet aroma when crushed between the fingers.

Dusty rose-colored flowers and silvery foliage make silky prairie clover (*Dalea villosa*) a real eye-catcher June through August on the sandy prairies of the steppe. Its habit is similar to purple prairie clover (*D. purpurea*), yet its stems are more flexible, lending it a graceful arching form. Its affinity for sandy soils and sharp drainage has probably limited its use in cultivation, but I do think it is worth the effort in a well-drained section of the rock garden.

Like many legumes, goat's rue (*Tephrosia virginiana*) disappears quickly under intense grazing, but where you do find it, it can be spectacular in late May and June. It has some of the most interesting flowers of all the legumes of the central steppe, usually banner-yellow outside, white inside, and the wings and keel rose-colored; the inflorescence is very showy on terminal racemes and can be up

The interesting tri-colored flowers of goat's rue.

to 4 inches long. Alternate, oddly pinnate compound leaves contibute to its airy appearance. This is another unique plant—again, probably limited in cultivation by its affinity for sandy soils but well-suited to the well-drained rock garden or xeric border.

IRIDACEAE

One of the rarest of all bulbs found in the central steppe, the stunning iridescent blue *Nemastylis geminiflora* (prairie iris) flowers in April and May in moist to drier prairies. From a tough globose bulb often 6 inches deep even in the hard clay, two or three slender grasslike leaves arise, followed by two or three flowers, 2 inches wide and held 6 inches in height. The pods, packed with reddish brown seeds, are reminiscent of a short, blunted iris capsule. The first time I found a colony of these wonderful bulbs in full bloom, I was stunned by their dainty disposition, and the absolute beauty of their six-petaled flowers made this a lifelong love affair for me. This unique bulb deserves to be grown in every garden, and although the bulbs can be grown from seed, with patience, they are still very hard to come by, given the several years the bulb needs to mature and flower.

LAMIACEAE

Scutellaria resinosa (resinous skullcap) is a low, stiff, extremely tough subshrub that makes a dazzling display of purplish blue splotches across the steppe, April through May. Rarely exceeding 16 inches tall, many erect stems arise from a woody base, with small, egg-shaped, opposite leaves arranged along their length.

Nemastylis geminiflora.

Yellow sundrops thriving
in a difficult spot!

LEFT *Scutellaria resinosa*
'Smoky Hills' is a
choice selection.

Plants can be covered with the two-lipped flowers, which spring from nodes in the upper part of the stem. The "skullcap" arises from the shape of the enlarged fruiting calyx, which produces four tiny seeds enclosed in a black nutlet. Resinous skullcap is a great addition to the rock garden or the well-drained, xeric border in strong sun. Without sharp drainage, it will die out quickly. I grew one specimen for over ten years on a harsh south slope at 6700-foot elevation in Colorado; it bloomed every year and never received supplemental water.

ONAGRACEAE

Calylophus serrulatus (yellow sundrops), a wide-ranging species with a variable morphology, is a plant that should be in every garden. It is an erect multistemmed subshrub to 2 feet tall in the eastern and northern part of its range, a more prostrate 8-inch spreading form in the south and west. Its major attribute is the abundance of clear, yellow, evening primrose flowers, only an inch wide. Peak bloom is usually May and June, but I have seen flowers on it nearly every month from spring through frost, if moisture is adequate. Tough and adaptable to a wide range of growing conditions in cultivation, and a real must-have for the mesic to xeric border.

Oenothera macrocarpa (Missouri evening primrose) has many natural variations in leaf shape, foliage color, flower size, petiole color, and habit—all represented by different subspecies on the central steppe. One can see all the subspecies, varying with the geography, in a long day driving around Kansas. Subspecies *macrocarpa*, the type, pairs its large vespertine flowers with dark green, polished

PLANT PRIMER

Missouri evening primrose in the garden, late May.

RIGHT Hoary evening primrose growing out of limestone at Clark State Lake in southwest Kansas.

leaves; large four-winged seed capsules follow the bloom and are designed to be pushed across the grasslands with the fall winds, blithely spreading seeds as they travel. It is found in nursery catalogs throughout much of the cultivated world and is a not-to-be-ignored favorite in the perennial border. Unfortunately, Missouri evening primrose is prone to severe depredation by flea beetles, particularly during the hot, dry periods so often experienced in recent years in the U.S. West.

Oenothera macrocarpa ssp. *incana* (hoary evening primrose) is abundant but localized on dry hillsides and rocky limestone promontories in grassland, ranging from the near eastern edge of the steppe of central Kansas well southwest to drier mixed and shortgrass prairies of southwest Kansas, western Oklahoma, the Texas Panhandle, and eastern New Mexico. It differs from the type in that it has very hoary, pubescent, distinctly blue-colored leaves; it has about the same size flower and seed capsule as the aforementioned, but overall plants tend to be a bit smaller in circumference. 'Silver Blade', an extremely silver-blue plant released by Plant Select, is the best foliage selection of this vibrant prairie perennial.

Oenothera macrocarpa ssp. *oklahomensis* (Oklahoma evening primrose) differs from the straight species in that it has elongated, dark red petioles and tends to be more prostrate than the other evening primrose subspecies. The deep color of the petioles coupled with the dark green shiny foliage help the clear yellow flowers stand out even more, if that is possible. 'Comanche Campfire', a Harlan Hamernik introduction, is the best selection of the subspecies on the market.

Oenothera macrocarpa ssp. *fremontii* is probably the most xeric of the evening primroses. This delightful, somewhat more diminutive subspecies has the blue-green foliage of *O. m.* ssp. *incana* but smaller flowers and seed capsules and linear

ABOVE 'Comanche Campfire',
from prairie plantsman
Harlan Hamernik of
Bluebird Nursery.

ABOVE LEFT A mass of
'Silver Blade', DBG.

LEFT Fremont's evening
primrose, close-up of
foliage and flower.

'Shimmer', with extremely linear blue foliage.

leaves. Fremont's evening primrose is at the center of its diversity in the Smoky Hills of north-central Kansas, where it literally covers the limestone-strewn road cuts. I have seen forms in this area ranging from broad linear leaf and darker green coloration to a nearly grass-like leaf with the distinct blue coloration. The grassy blue form is well represented in 'Shimmer', a cultivar patented by Scott and Lauren Springer Ogden. 'Shimmer' is an excellent plant; not only can you still enjoy wonderful yellow blossoms but even the flowerless lacy form of its foliage is worthy of inclusion in any dry border design.

PLANTAGINACEAE

Penstemon secundiflorus (side-bells penstemon) is a very showy plant in full bloom, May through early June, with an abundance of bluish lavender flowers clustered along one side of the stem, which varies from 8 to 20 inches tall. Thick, blue-gray leaves clasp the stem, adding to the attractive appearance of the plant, which is a natural for the rock garden or a naturalistic xeric border. Or consider the unfinicky *P. tubaeflorus* (tube penstemon), a sight to behold in the eastern regions of the steppe, with clear white inflorescences of narrow interrupted panicles, upward of 30 inches tall. Common in hay meadows and open woodlands, this amazing penstemon grows in sand or clay, wet or dry, which makes it the perfect penstemon for those who lack the knack for growing the more finicky western species we all covet. I grew great masses of tube penstemon along the edge of a pond in the stickiest clay soil imaginable in eastern Kansas, and the plants thrived.

Cobaea penstemon flowers are some of the largest in the genus.

LEFT Tube penstemon growing in a hayfield in eastern Kansas.

But *Penstemon cobaea* is my absolute most favorite of the dazzling genus, without a doubt. A robust perennial to 3 feet tall, its massive flowers, mostly white or pink to light or dark purple, are an absolute treat for any penstemon lover. The tubular throats of the individual flowers are lined with dark red to brown lines—basically runway lights for bees and other pollinating insects. The beefy lanceolate to lance-ovate leaves are dark green and shiny, as if their surface had been buffed with wax. Late May and June, the robust flowers appear all over the rocky hillsides of the Flint Hills, providing a display I will never forget. If cobaea penstemon has one fault, it is that it tends to be short-lived—maybe because it puts so much energy into its huge flower displays. Try it in the hot, sunny border with good drainage.

POACEAE

Bouteloua gracilis (blue grama), a stately 2-foot-tall grass arising from a dense tuft, is widely used in gardens for its one-sided inflorescence, which gently curls as it matures to form an eyelash. 'Blonde Ambition', a Plant Select introduction, has a more clump-forming habit, and the seedhead emerges chartreuse, changing to blond as it ages. Blue grama maintains its appearance and distinctive seedheads throughout winter. It is often included in seed mixes with buffalograss, as it too tolerates frequent mowing. No garden in the American West, from xeric to more mesic situations, should be without this very adaptable and mesmerizing grass!

Muhlenbergia reverchonii (seep muhly) is a slender grass arising from dense, fibrous tufts reaching 1 to 3 feet tall. The delicate and airy seedheads are purplish

'Blonde Ambition' blue
grama, late summer, DBG.

and much branched. High Country Gardens introduced Undaunted, a ruby muhly grass with amber-pink flowering plumes, several years ago. It thrives in full sun and most any type of soil from sandy to clay, plus it will tolerate less-than-perfect drainage conditions and heat. Its spectacular airy mounds are a delight in late summer and fall in the mixed perennial border and even carry their good coloration well into early winter.

In the last 15 years, *Schizachyrium scoparium* (little bluestem), one of the most widespread grasses of the central steppe, has become a garden staple throughout the United States. Its erect habit (up to 36 inches in height), varying foliage color (light green to dark powder-blue), and wispy seedheads make it an easy selection for the perennial border. Not particular about soils, it prefers slightly dry to mesic growing conditions to reach its full potential; but it is a warm-season grass, so in

Little bluestem adds color and
structure to the winter garden.

Clematis hirsutissima var. *scottii*
on a dry slope in Colorado.

colder climates it is well into late spring before it rouses from dormancy. Many cultivars are available, including 'Blaze' (known for its attractive amber fall color) and 'Prairie Blues' (gray-blue during the growing season, dark purplish fall color).

RANUNCULACEAE

Clematis fremontii, a non-vining clematis endemic to the Great Plains, is truly a unique and long-lived specimen for rock gardens or near the front of the perennial border. Reaching only 10 to 12 inches tall, numerous stems arise from a tough, woody base, supporting ovoid, leather-textured, dark green leaves. In April and May, light blue to gray-violet upside-down urn-shaped flowers appear, with recurved margins more yellowish to light cream. Soon after, clusters of short-plumed seeds arise and persist well into the fall. Fremont's clematis has a coarse fibrous root system that anchors very deeply in the soil, but it still can be divided and moved around the garden if care is taken in the fall to protect it from winter desiccation.

Another wonderful and mesmerizing non-vining clematis from the dry grasslands and higher montane steppes, *Clematis hirsutissima* var. *scottii* (Scott's sugarbowl) produces mounds of lacy, blue-green foliage up to 12 inches tall, followed later in May and June by nodding blue flowers with a dense covering of fine hairs on the exterior. Wild tufts of golden seedheads follow and remain through much of the growing season, adding to the season-long interest. Prefers a dry sand to gravel but is tolerant of more clay soils. This taxon was released (as *C. scottii*) by Plant Select in 2013. A must-have for any serious rock gardener.

Prairie Gold aspen, *Populus tremuloides* 'NE Arb'.

SALICACEAE

Populus tremuloides (quaking aspen) is ever-loved for its shimmering golden leaves in fall, its striking white bark, and the flat petioles that allow its leaves to rustle distinctively in the slightest breeze. Many a steppe resident has lusted after this tree after a trip to the Rocky Mountains, but alas, the heat, humidity, or sheer aridity of the plains had foiled the best attempts to reap its wonderful attributes. Even along Colorado's Front Range communities, quaking aspen suffers from heat stress and numerous bark cankers and leaf diseases. Enter 'NE Arb', an introduction from Nebraska's GreatPlants program. The original tree, found in east-central Nebraska, was clearly able to withstand the heat and humidity that had been the undoing of the species in much of its (attempted!) cultivated area. This strong-growing selection reaches 30 to 40 feet with an equal spread and has light green to white bark. I have grown it in the Denver area for several years; my only caution is that the vivid "prairie gold" of the high-mountain taxon may not develop in the lower regions of the plains and eastward, but all the other attributes of the species are present and accounted for.

SAPINDACEAE

The first time I spotted *Sapindus drummondii* (soapberry) growing along a stream in the Red Hills of Kansas, I could not understand why this worthwhile and striking tree is not more widely cultivated. Its alternate and even-pinnate leaves

Soapberry in fall color along a creek in the Texas Panhandle.

give it a soft and graceful appearance, the white flowers are not overly showy but still pleasing, and the globose, translucent yellow-orange berries are captivating. What is not to like about this tree, especially in autumn, when groves of it shimmer a deep yellow out on the plains! Reaching 40 to 50 feet tall and wide with a vase to rounded habit, this tough tree—which is not finicky about soils, drainage, or air pollution—deserves wider cultivation as a street tree and in park settings. Unfortunately, we have become so adverse to messy fruits falling on our precious grass and sidewalks, we may never fully embrace this wonderful inhabitant of the steppe.

PLANT PRIMER

THE INTERMOUNTAIN NORTH AMERICAN STEPPE

DAN JOHNSON

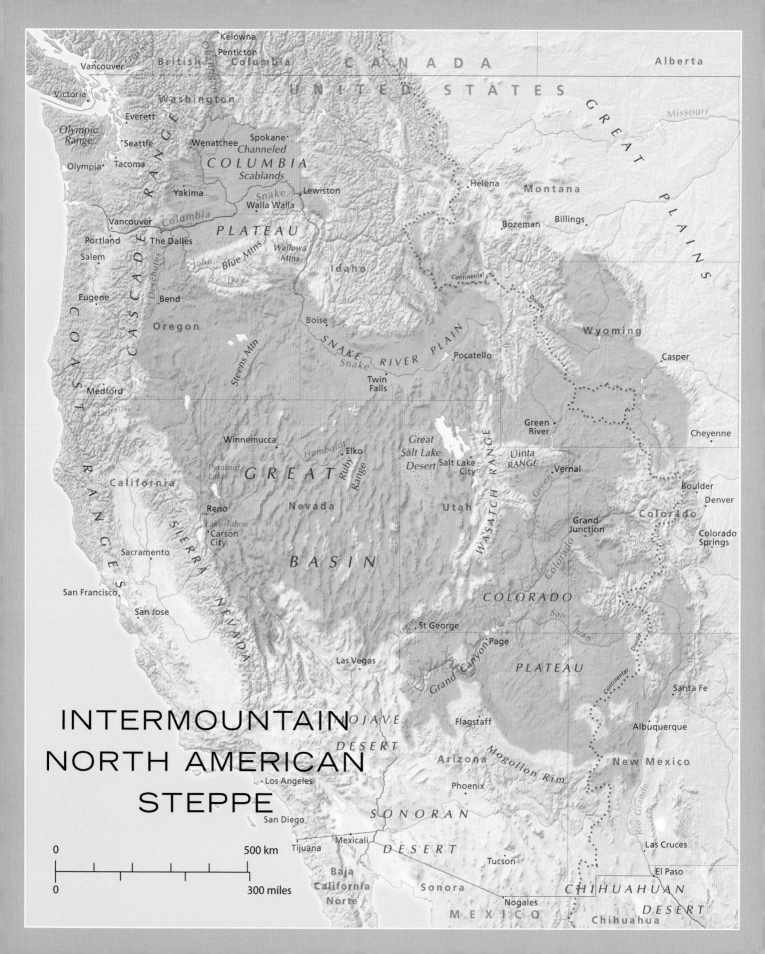

INTERMOUNTAIN
NORTH AMERICAN
STEPPE

Geography and Climate

Like all steppe regions, those in North America are the product of complex climatic processes that often begin hundreds or thousands of miles distant. Subtle shifts elsewhere may send moisture skimming by to the north or south, or blocking areas of high pressure or typhoons in the tropics may send storm tracks reeling temporarily off-kilter. Since precipitation is inherently low to begin with, these shifts can set in motion the extremes of drought and deluge that shape the landscape, the flora, and the cultures of the steppe.

Weather systems typically enter the central portions of North America from the west, but given this pattern alone, most of the western United States would be maritime forest. The cool moist air of the Pacific would stream unimpeded over the North American continent, blanketing the valleys in deep forests sequined with massive lakes and rivers.

But there's another major player in this scenario. The east front of the Pacific Ring of Fire has rumpled the west coast of the Americas in a nearly continuous 12,000-mile mountain strand, from Alaska's Aleutian Islands to the tip of Patagonia. The ranges run generally north to south, often in abrupt opposition to prevailing westerly winds. If these mountain ranges were less continuous or bent significantly east or west, we would live in a very different world, but with its current configuration, the classic rain shadow is the result. Near Vancouver, B.C., the Coast Range of Vancouver Island and the northern Cascade Range draw vast amounts of water from heavy clouds forced up their forested slopes. To the south, the jagged Olympic peaks take the brunt of any storm, creating a small rain shadow of their own before the storms cross inland. The seaward Hoh Rainforest fairly swims in up to 170 inches of rain a year, while leeward Port Townsend is

Sage, piñon pine, and long vistas characterize the Intermountain West.

comparatively crisp at just over 18 inches—a scenario that is echoed in varying permutations all the way to Tierra del Fuego. This western mountain boundary sets the stage for all that lies beyond, and east of here, the world runs dry. Climate zones of North and South America follow this general model, though local topography and oceans make the reality much more complex.

The intermountain grass and sagebrush steppes occupy a vaguely elliptical region of the Interior West. Their western border is well defined by the rugged Coast and Cascade ranges, merging to the south with the shrouded summits of the Klamath range and continuous ramparts of the Sierra Nevada. This great wall of mountains creates sharp and predictable gradients in precipitation, marking the western edge of two significant divisions of the intermountain steppe: the Columbia Plateau and the Great Basin.

Throughout the Columbia Plateau and the western Great Basin, winter-precipitation regimes favor shrub communities over grasslands, which appreciate summer moisture. This is often what tips the scales from one plant community to the other, and these communities can shift over time as climate patterns shift. This gradient changes as one moves east. As the likelihood of summer thunderstorms

Grasslands spill through the Southern Rockies of New Mexico.

increases, so does the proportion of grasslands. This is due in large part to the North American Monsoon, a generally predictable shift in weather patterns caused by high pressure systems that dominate Texas and the Southern Plains in midsummer. This swings waves of moisture up from the Gulf of Mexico and even the Gulf of California into the Southern Rockies and the deserts of the Southwest. Strong pulses can result in drenching desert storms and flash floods, especially in the southern and eastern portions of the region.

The steppe's eastern edge is generally defined by the Rocky Mountains, but this is by no means a clear singular line. In some areas the division is complete, a wall of peaks dense with aspen and spruce. In most areas though, tendrils of steppe wind their way up low valleys, or occupy grassy south slopes, with northern exposures shrouded under sweeping Douglas fir. In several places the mountains give way altogether, and intermountain steppe transitions into Great Plains prairie or Chihuahuan grasslands. Widely separated ranges can frame distinct regions, as happens with the Colorado Plateau, bound between the Wasatch and Uinta ranges and the Southern Rockies, and to the south, falling away off the forested Mogollon Rim into transitional woodlands and true Sonoran Desert.

Fire also plays a deciding role in these plant communities. At least five species or subspecies of shrubby artemisias occupy various habitats in the steppe, but the sagebrush communities are not fire-dependent, and actually regenerate slowly after fire. Grasslands, on the other hand, flourish when faced with periodic fires and may return more quickly to burned sagelands than the shrub community does. The unfortunate addition of highly flammable cheatgrass (*Bromus tectorum*) has turned some of these processes on their head, invading both communities rapidly after fire and making areas even more fire-prone. This is one of the great challenges to rangeland management in the West.

Sagebrush steppe still dominates millions of acres in the West, and much of it is still reasonably intact. A wealth of wildlife depends on its continuation, and countless wildflowers and primal vistas still await anyone who yearns for solitude in the timeless open country of the Intermountain West.

COLUMBIA PLATEAU

The central plateau of the Columbia River watershed, the most northerly of the intermountain steppe regions, is a vast landscape of contrast and high drama. Its volcanic cones, coastal forests, rain-shadow patterns, basalt-capped mesas, and dominant shrublands are nearly a mirror image of what happens at the same latitude south, in Patagonia. The higher peaks of the southern Andes capture more snow, producing active glaciers and ice sheets, and the proximity of the Atlantic to the east also moderates temperatures when compared to the vast continental effect of North America, but the commonalities are undeniable.

While tropical storms at lower latitudes move west away from Mexico, moisture-laden storms in the North Pacific have more than 5000 miles of open water to swirl and morph eastward from Japan toward the Gulf of Alaska and the Pacific Northwest. In the depths of winter, as the storm track sinks southward, so do the storms, resulting in the winter rains characteristic of California's Mediterranean climate. This climate exists on a south-to-north cline, resulting in what amounts to a progressively cooler Mediterranean climate as you venture north along the coast into Oregon, Washington, and even the southwest corner of British Columbia. The results are generally wet winters and dry summers.

In most years a steady procession of winter storms are brought up short by the Coast Ranges. "Brought up short" may be overstating it, since there are places where the Coast Range barely tops 1000 feet. In the grand scheme of things, that's the equivalent of a pencil lying at one end of a football field—barely a ripple—swamped by a skiff of atmosphere. Still, that is enough to squeeze out substantial rainfall, and where the mountains rise to 5000 feet and more, they drip with moss, ferns, and towering forest giants of red cedar, spruce, and fir. Some rains never make it beyond these green velvet ridges, and on their leeward side, vegetation thins into grassy pine or oak savannas at lower elevations. But the strongest winter storms press onward, drenching the farms and towns of the Willamette, Rogue, and Sacramento valleys in agricultural bliss.

The great Sierra Nevada create an abrupt rain shadow.

The next obstacles are more formidable. Storms partially spent in the Coast Ranges are lifted higher, driving gray swells of mist into the icy teeth of the Cascades, the Klamath, and the Sierra Nevada. These are the game changers, wringing the clouds dry as they lift to 7000 feet, cresting the likes of Mount Hood (over 11,000 feet) and Rainier and Shasta (over 14,000 feet). As expected, these ranges capture most of the precipitation and return it west into the Pacific, the eastern slopes left high and dry.

From the Alaskan archipelago nearly 2000 miles to Vancouver, B.C., the mountains generate their own glaciers and countless rivers that cataract their way to the sea, but not a single river crosses this range from the interior. Only at Vancouver does the Fraser River wend its way from drier interior climes to the Strait of Georgia and the Pacific. Pacific moisture pushes up the valleys, densely wooded from ridge to shore. The parallel Cascade and Columbia mountains arc in unison, about 150 miles apart but essentially connected by a rugged, forested interior plateau. But some 200 miles from the Pacific, the headwaters of the Fraser, Thompson, and Okanagan rivers are tucked into the rain shadow of the Cascades, the valley air deprived of moisture. Here the first threads of interior steppe are found, and a landscape more like Colorado emerges, complete with sagebrush and tufted grasses dotting the dry slopes. Here, a thousand miles north of Colorado, the steppe appears at just 1000 feet above sea level, creeping up the slopes to 3000 feet before the Northwest forests close in again.

This pattern reappears in each valley and patchworks its way up and down southern exposures for some 100 miles to the north and east, up the Fraser River to Farwell Canyon, up the Thompson River to Kamloops, and down the Nicola Valley to Merritt's grasslands. Kamloops manages on just 8.5 inches of precipitation a year. The small town of Merritt receives just 12 inches, while Hope, just 60 miles southwest, slogs about with 79 inches!

Another short jump to the east, Vernon, Penticton, and Osoyoos trace the Okanagan Valley south toward Washington state. This narrow "banana belt" of Canada can break 100°F during its warm summer months, its low broad river valley growing ever wider as it bears south though ancient hills, scablands, and lake basins, coursing south to join the great Columbia.

The Columbia existed long before the Cascades, carving its spectacular gorge as the mountains swelled beneath it, the largest of the three rivers on the entire coast to leave the interior steppes and reach the Pacific. The Columbia Gorge is a reliably windy place, inland high pressure pushing west toward the sea, or rising hot interior air sucking in cool breezes from the west.

The climate of the Columbia Plateau itself is classic shrub steppe, with cold winters and hot summers, scant rainfall, and long horizons. The rain shadow cast by the Cascades limits precipitation, but unlike much of the intermountain region, winter clouds and fog can be frequent phenomena. Precipitation is primarily a winter event in the form of light rain or snow. Higher humidity chills to the bone. By early summer the clouds have cleared and the temperatures warm steadily to desert-like conditions. Deep-rooted shrubs tough it out, but all else dries to a crisp in the long summer days. While winter lows have occasionally been recorded in the −30°F range or lower, highs climb into the low 100s each summer, with record highs of 118°F.

GREAT BASIN

Conditions vary greatly between snow-capped peaks, forested ravines, salt playas, sand dunes, sheer vaulting mountain ranges, and broad valleys filled with boulders and alluvial soil. This topography is comparable to large portions of Central Asia, with rock-strewn mountain ranges separated by austere valleys. Broad alluvial fans, sage shrubs, bunchgrasses, dark junipers dotting rocky slopes, snowy peaks on the horizon—these are hallmarks of the Great Basin steppe.

The boundaries of the Great Basin can be defined in several ways, leading to some confusion. Defined strictly as a watershed, the Great Basin includes portions of Oregon, Idaho, California, Nevada, and Utah. It is distinct only in that it acts as a broad bowl that waters may flow into but will never escape. There is no connection—other than atmosphere—to the sea or any other water system of

Lupine and sage are a common combination in the sagebrush steppes, softening the landscape.

the world. Water drains and pools into wide valleys, only to sink into aquifers, or it evaporates, leaving its assorted salts and minerals behind in chalky playas.

Defined by geology, the term "basin and range" refers to a physiographic expression of unique geologic features—put simply, a repeated series of uplifted fault block mountain ranges and intervening dropped valleys (known as a horst and graben landscape) caused by the immeasurable forces of plate tectonics pulling the continent apart. This pattern extends far beyond the sagebrush steppes, however, stretching through southern Arizona and New Mexico, along Mexico's Gulf of California and deep into central Chihuahua. In some areas, these isolated mountain uplifts are known as sky islands.

Defined as a floristic province, the combined regions known as the Northern and Southern Great Basin and the Snake River Plain most accurately describe the sagebrush biome that we will collectively refer to as the Great Basin.

Throughout the Great Basin, dry bajadas slope away from scoured canyons with geometric precision, intersecting at the low point with corresponding bajadas from the next mountain range. Much of this country bears little resemblance to the sagebrush steppe. In fact, some sources paint it all with the broad brush of "Great Basin Desert," but that's an inaccurate simplification.

The basin as a whole has a generally convex profile from west to east. The harsh Lahontan Basin occupies the lower western region and the larger Bonneville Basin occupies the lower eastern section. The central basin has a higher average elevation than either edge and is punctuated by the even higher calcareous mountain ranges. This is far from a barren wasteland. On each mountain range (and there are hundreds), between salty playa and the alpine, is a pyramid

Ragged mesas and red rock
of Capitol Reef contrast
with the Henry Mountains.

of life zones. Depending on elevation, each is ringed in the appropriate halo of saltbush, sage, or rabbitbrush, bunchgrass, piñon and juniper, pine and fir, and if tall enough, a cap of alpine tundra.

In winter, Pacific storms roll onto the west coast, giving California its Mediterranean-style winter rain cycle. The bit of winter moisture that manages to squeak over the Sierras delivers the western portion of the basin much of its precipitation as snow. Higher elevations see the highest accumulation, which percolates down into porous soils to feed small mountain springs and streams. Beyond the peaks, storms thin into snow squalls that bluster across the empty basins where dry winds can evaporate the snow before it has a chance to melt into the ground—melting "from the top up" as plantsman and author Bob Nold would say—a process known as sublimation. Drying winds accelerate this process, and this phenomenon is common throughout the arid Intermountain West. The basin and range topography also lends itself to temperature inversions as cold air pools in the many valleys. Heat radiates away quickly in the dry air, and winter lows are frequently below 0°F, with the all-time record at −50°F.

When strong low pressure systems settle in the basin, pressure gradients can produce easterly downsloping winds that buffet cities along the western Wasatch Front. When high pressure settles over the Great Basin, weather is sunny, calm, and dry. If conditions are right, winds push outward at the fringes, toasting southern California in the autumn with their famous Santa Ana winds, or drying downslope winds east of the Wasatch Range.

The eastern portion of the basin sometimes benefits from monsoonal summer thunderstorms as moist air nudges up against the Wasatch Front, but this is not reliable. Drought-like conditions are the norm here, the lower elevations just a storm or two away from shifting into true desert.

COLORADO PLATEAU

In Utah, the parched earth splits open in a chaos of red, tan, and ochre sandstone chasms draining a great bowl of distant mountains: the Wasatch to the west, Uinta to the north, Southern Rockies to the east. All are high enough to interfere with approaching weather systems, creating a rain shadow effect on three sides. Still, they gather reliable snow and rain, filling the rivers and streams that dissect the plateaus below, creating rich and unique habitats throughout the basin. Terraced like a grand amphitheater, layers of sedimentary rock direct all runoff toward the Colorado River and lower elevations to the southwest. The uplands are remote and unpopulated, and little explored by most. But the lands below this are a recreational paradise, with most land publicly owned and more national parks and monuments than any other part of the United States.

The canyon country is a dynamic place, and weather can transform it in every

Water-etched cliffs near
Utah's Fremont River.

way. Fog can rest in deep ravines, or clouds may envelop red stone towers. Sunlight may flash on a canyon wall framed by purple-gray clouds. Calm morning light clear as crystal can set every stone and shadow on edge, until midday glare flattens the world and rounds the corners off every wall and monument.

The Colorado Plateau experiences two potential seasons of moisture each year, as compared to the winter-precipitation patterns of the Columbia Plateau and the Great Basin. The northwest portion of the plateau is generally the driest and coldest region. There is a gradual transition here as one travels southeast, the climate shifting from that of the Great Basin to the more temperate canyon country. Elevation and latitude both play a part. Winter nights are frigid, but days are often mild. Light winter snow may outline every rocky ledge like a fine etching. This melts away quickly on sunny exposures but may linger all winter in the shady depths of deep ice-box canyons. Passing summer or winter sprinkles may briefly pepper the red rock before being sucked away by the porous sandstone. Summer highs may easily reach 100°F or more, especially in the southern canyons. These same locations may see extremes of −15°F in the depths of winter, but in the northern reaches of the plateau, lows have fallen to the −40°F range.

Summer moisture is more likely here thanks to the northward sloshing of the North American Monsoon storms each July through September. This is still a high desert/shrub steppe, with low and mid-elevations receiving just 8 to 15 inches of precipitation, but the summer rains can create a distinctive late summer and fall flush of greenery and flowers not typically found farther north and west. Entering from the south, moist subtropical air streams into Arizona, the orographic lift generating plentiful storms that drench the ponderosa and fir forests of the sky islands and Mogollon Rim. If the flow is strong enough, it surges on, lifting again to form crisp-cut thunderheads above the mesas and canyons. These summer thunderstorms provide most of the plateau's moisture. They are a common event in this country and one of the primary sculpting forces at work on the soft sandstone. Storms can be sudden and severe, sending rusty chocolate torrents bursting over cliffs and down narrow ravines, pushing rocks and pungent debris on ahead. A deluge in faraway mountains can travel down dozens of miles of dry canyon washes before its energy is spent or it joins a permanent stream, and years may pass between such events. This is the great hallmark of the Colorado Plateau: a world deprived of water yet shaped in every way by it.

Touring the Intermountain West

FOLLOWING THE COLUMBIA

The Cascade Mountains seem benign, the fragrant orchards and quilted farms of the Hood River Valley oblivious to the slumbering giant of Mount Hood. But the Cascades are alive, and their great ridges, softened by red cedar, fir, and rain, are the perfect disguise for the forces that shaped this region. Frosted silver peaks claw the horizon; miles-wide craters drop the floor out of the forest; still-as-death basalt flows smother half a valley. These volcanic punctuations speak of the region's true history, and east of the Cascade Crest, lava abounds.

To follow the Columbia River inland from the Pacific is to read a textbook example of the steppe. The river itself has long since been tamed by hydroelectric and flood-control projects, its rapids smoothed and buried shore to shore, but basalt walls still frame its course. This fabled river conjures certain images for those who don't know it, of dripping evergreens and shoreline cabins tucked along forested coves. This is an accurate image, and for nearly 150 miles inland it passes through exactly the lush forests we all imagine. East of Troutdale dense forests cover every hill, and gray walls split open to reveal storybook waterfalls plunging Rivendell-style from lichen-painted cliffs. Latourell, Shepherd's Dell, Bridal Veil, Multnomah, Oneonta, and countless others bathe the glens in perpetual mist, and moss drapes anything that does not move, as lush as any tropic paradise.

But here, with just a few bends in the road east of White Salmon, Washington, the world changes. Ravines and west-facing hills buffeted with mist are green with forest, east- and south-facing slopes grassy and parched by interior winds. Across the river at The Dalles, wine-tinted cheatgrass ripples like wheat up broad volcanic slopes. By the time you reach the Deschutes River, the transition is complete, brown mounded hills and bluffs on every side. For the next 500 miles the great Columbia gnaws at the deep dry bones of the Northwest.

The Columbia Basin is vast, even larger than the drainage of the Colorado River, and carries more water to the Pacific than any other western river. All along the east Cascade slope, the forests give way gradually as the land slowly drops from under them. Between 2000 and 4000 feet the forests thin, ponderosa pine dipping into waves of grass, and sage cresting the ridges. The sagelands are enormous and diverse, with at least five species of sage mingling with bunchgrass meadows and army-green stands of *Juniperus occidentalis*. This basin boasts as violent a geologic history as any on earth. Only its high edges are softened by the green veil of forest, its interior windblown and parched as desert, scratched and laid bare. Basalt flows, ancient lakebeds, tortuous river gorges, layered sediments from glacial floods, rolling grasslands—there is little lacking here—the whole of it once clothed in sage and bunchgrass or black lava.

Broad floodplains border the Columbia as you continue east, and though near to the river, the conditions change quickly. Midwest-style valleys dotted green with pivot irrigation reach the hazy horizon, producing everything from melons and wine grapes to potatoes. Winters are cool and overcast, but summers are scorching, hot enough to sweeten miles of Walla Walla onions, plump for harvest by mid-June.

The Columbia turns north at Wallula Gap, the only place where the river was able to escape the interior basin. It's joined by the Snake River near Pasco, and both rivers tell long and repetitive geologic tales of melting glaciers, catastrophic Pleistocene floods, and sprawling lakes, slowly deepening the gap on their way to the Pacific. These ancient lakebeds were filled with sediment time after time as failing ice dams sent inconceivable floodwaters raging from far upstream, eventually draining through the Wallula and backing up the Willamette Valley before finding escape to the sea. For over a hundred miles upriver these lakebeds stretch east, now a luxuriant green with farms, the Columbia cutting its way still deeper into the volcanic bedrock.

West of the river the land tilts up toward the east Cascades, the forest edge zigs and zags northeast. Corrugated sage hills frame sunny valleys rich with deep volcanic soils. The towns of Toppenish, Yakima, and Ellensburg go about their agricultural business below the distant dome of Rainier. Irrigation long ago transformed the place. The rain shadow rules, at only 8 to 10 inches of precipitation—about the same as Phoenix—but since the mid-1800s, agriculture has blossomed on over half a million acres of apples, vineyards, and lush fields of mint and hay.

At Trinidad the landscape inverts itself. Valley farms are left behind, and 2000 feet above the river, miles-wide plateaus open into checkerboards of wheat, the original sagebrush steppe but a century-old memory. Where wide shores allow, farms hug the river, but most of its course is bedded in solid rock, basalt mesas splayed in all directions, edges chipped away high above the water. Known as the loess islands, these deep layers of basalt—up to 2 miles thick—are capped with ancient loess soil deposited by wind over the eons. The plateaus form an archipelago of highlands carved apart, and agriculture has flourished on these fertile soils, every arable corner long since turned by the plow. The percentage of native vegetation that remains is in the low single digits at best, relegated to the rough edges of these intensely cultivated farms.

It is easy to find similarities here with the dry plateaus of Patagonia roughly 7000 miles to the south. A quick visual of either locale might be hard to distinguish. Windswept lava plateaus clothed in shrubs, volcanic peaks sharpening the horizon, tufted grasses bending to dry winds. At a comparable latitude, similar geography contributes to similar weather patterns shaping both these regions into steppe, though the continental mass of North America alters the effect in some ways. Rivers emanating from the interior allow for more exploitation of the land in North America, and irrigated farms abound. Much of Patagonia remains remote and uncultivated. Its comparatively narrow spit of land is flanked by two oceans producing incessant summer winds that preclude most agriculture in the bulk of Argentina's steppe, but the mix of shrubs and bunchgrasses is similar. One common adaptation to the summer winds of Patagonia is the prevalence of cushion plants, which stay low and warm in the face of chilly winds that hover in the 40- to 60-mph range for weeks on end. This growth form is less common in the North American steppes.

Northeast of the loess islands, the foothills of the Northern Rockies outline a fuzzy boundary as fingers of steppe and forest drift southeast from the Okonagan Valley to Spokane, and south to Lewiston/Clarkson. The icy headwaters of the Columbia tumble out of the Canadian Rockies coursing north at first, then south through the chilled forests of the high country. But somewhere a few miles north of Davenport, Washington, and tamed beneath the waters of Lake Roosevelt,

Wild buckwheats (*Eriogonum compositum*) brighten a slope underpinned with dark basalt.

the river slips quietly past its tree-lined banks, into the drier hills and sagelands of the steppe.

The waters we now call the Columbia have not always stayed within these banks, and just beyond its current course are signs of something big. The farmed plateaus are widely spaced, and the scoured landscape left between them, known as the Channeled Scablands, was a source of mystery and contention for many years. Potholes, dry falls, braided coulees (dramatic—but often dry—watercourses), and patterned swells of land made little sense at eye level, but in the 1920s aerial photos prompted geologist J Harlan Bretz to string these clues together. Some 10,000 years ago and more, the Ice Age ebb and flow of the Cordilleran Ice Sheet to the north periodically blocked drainages and created the vast lakes Missoula and Columbia. Eventually conditions would shift, and weakened ice dams at last gave way, discharging the entire lake in a matter of days in apocalyptic floods. Waterborne boulders chewed up cliffs and drilled into bedrock. Rinse and repeat every 40 to 60 years. Soils were stripped down to bare basalt and carried off to settle in ancient lake basins, now the farmlands of Quincy, George, and Royal City. This happened enough times and with such violence that eroded channels

OVERLEAF Flower-strewn meadows in the highlands near Pendleton, Oregon.

and plunging waterfall cliffs look like a landscape in waiting, just add water. By the time our inquiring minds arrived on the scene, all was quiet.

These scablands are rugged, their soils often thin with basalt ribs and elbows poking out, no place for farming. Grass and sage still dominate a landscape suitable for desperate grazing and inquisitive exploration into an unnerving past. Vegetation cover is typically thin, though impermeable rock can pool seasonal rains into sparkling lakes and lush wetlands. The land is open and untrammeled, home to abundant wildlife, and wet winters and springs bring a flourish of balsamroot, mariposa lilies, wild buckwheats, camas, shooting stars, wild onions, ball cactus, penstemons, lupines, and mounding phlox. Most of the coulees carry little more than streams at best, but a few boast respectable features that conjure the tempestuous past. The improbable Palouse Falls bursts through its black and brown canyon and leaps off 180 feet of sheer basalt into a vast plunge pool formed by floods many times its current size.

Just southeast of the scablands, farms reappear on a rippled landscape of ancient dunes, the photogenic Palouse Prairie and, further on, the Camas Prairie beyond the Clearwater River. Silt blown in from ancient disturbances to the west accumulated here in an expansive undulating sea of hills and swales. The prairies were once one of our richest troves of wildflowers, nearly lost now in an ocean of swaying wheat. A few remnants remain, and in the mix are strong ties to the meadow wildflowers of the nearby Northern Rockies, replete with Idaho fescue, paintbrush, prairie smoke, larkspur, mule's ears, iris, helianthus, and dozens more, all adrift on billowing emerald hills.

To continue our circuit of the Columbia Plateau, we now must turn west. A comparatively unknown range of hills and peaks blocks our way south. Treasured by locals, Oregon's Blue and Wallowa Mountains would be on a "best kept secret" list of places to spend time. But for the pleated chasm of Hell's Canyon, the forests of the Rockies at Seven Devils Wilderness would merge into these mountains. The Snake River effectively carves them apart with America's deepest canyon—nearly 8000 feet top to bottom. The history of the Snake mirrors that of the Columbia, rife with cataclysm and course-changing events. Though part of the Columbia watershed, its true roots are in the Middle Rockies and most of its length is spent carving its way through the northeast edge of the Great Basin beyond the Blue Mountains.

The Wallowas, sometimes called the Little Rockies, boast snowy peaks over 9000 feet, with lower slopes wreathed predictably in sage and grasslands. The Blues rise near Lewiston/Clarkson, corrugated ridges between 3000 and 6000 feet expanding to the southwest. Together, they are high enough to capture plenty of maritime moisture, especially winter rain and snow, and they share vegetation with both the Rockies and the Cascades. Ponderosa and lodgepole pine mingle with grand fir, western larch, and Douglas fir in lush forests, receiving as much as 100 inches of precipitation a year, compared to just 10 inches at nearby Baker City.

Sinuous arms of sagebrush steppe extend up each valley, edges indistinct as grassy pine parklands transition into thick forest. The many tributaries of the John Day River seep out of these hills and run north through tilted canyons of the Umatilla Plateau to the Columbia. Not far west, the Deschutes River runs parallel, fed by the east slope of the Cascades.

The Umatilla Plateau tilts north off the Blue Mountains, its upper slopes rumpled as cabbage leaves and wild as anywhere. The land levels out though, underpinned by thick basalt flows, true to the regional theme. Loess deposits support broad-swelling bluebunch grasslands, and plenty of dryland farming where conditions allow. Thinner soils bake dry in summer, supporting crisp grasses and sage.

The steady gnaw of rivers has unraveled much of this area's ancient history. The renowned John Day Fossil Beds are a window into an incomprehensible past, resting silently beneath grassy hills. Remains of humid tropical forests as old as 44 million years are preserved in massive volcanic ash flows with more than 170 ancient species identified, from ancestral magnolias and palms to rhinos and crocodiles, even ancient insects. Colorful layer upon layer reads like the pages of science fiction as the climate became cooler and drier. And all this is capped by more "recent" Columbia River Basalt flows, themselves over 15 million years old. Subsequent ash flows from the rising Cascade Range preserved mastodons and a world of more familiar horses, dogs, and lions in their ancestral forms, still 7 million years removed from our world.

The Deschutes River is born out of snowfields and forest lakes in the rugged central Cascades. The river passes through Bend on its high plateau, ice-crusted peaks dominating the skyline a mile above the town. A midtown cinder cone spiraled with a narrow road offers an impressive vista. Traveling through Bend one summer, a cold front left frost on my windows on 1 July, and crossing the Cascades, rhododendrons near McKenzie Pass flowered under 3 inches of wet snow.

Volcanic activity is not all as ancient here as the great basalt flows of the Umatilla. Just south of Bend, the Newberry volcanic shield is fully plumbed with lava tubes, caves, and fissures. Trees engulfed by lava as recently as 6000 years ago have rotted out, leaving their hollow molds, as newer ponderosa forests await the next burial. The more recent Big Obsidian Flow oozed from the south caldera just 1500 years ago.

Coursing east, then north, the Deschutes and its tributaries leave the stately ponderosa country and slip lower into the olive-green of juniper thickets, then into the sage and bunchgrass hills. Canyons carve a path much like the John Day, eroded basalt columns slumping below stark plateaus, the snowcones of Jefferson and Hood chipping the horizon. Left to itself, the land is dry, firmly in the

Cascade rain shadow once again. The towns of Madras and Culver average less than 10 inches of precipitation a year, but arable flats nearby are a patchwork of irrigated fields. The river is not long—just 250 miles—but its presence is vital to the people of the region. It meanders through primal and desolate country, finally joining the Columbia near The Dalles. West of these valleys, the juniper and sage of the East Cascade foothills complete the circle, returning us to the western edge of the Columbia Basin steppes.

INTO THE GREAT BASIN

Northeast of Bend, Oregon, the Blue and Wallowa Mountains shed their melted snows in all directions. Most of it flows either down the John Day River, or toward the Columbia and Snake rivers to the north and east. South of the Blue Mountains the transition to the Great Basin is subtle, imperceptible slopes directing water to one fate or the other. Indeed, it is only the divide of these watersheds that distinguishes the Columbia Basin from the Great Basin.

To the east this divide follows a zigzag of high ground through southeast Oregon and southwest Idaho, swinging south into Nevada, but plant communities spill freely from one ridge and valley to the next, blurring the lines. The expansive lava plains of the Owyhee Uplands sprawl eastward, all this tilted northeast and draining into the deep canyons of the Snake River. Below the impounded waters of Brownlee Reservoir, the Snake slips away down the throat of Hell's Canyon to join the Columbia.

In southern Idaho the volcanic tabletop of the Snake River Plain crunches underfoot in a 400-mile arc, swinging widely around the southern slopes of Idaho's Sawtooth and Salmon ranges. Despite an overall uniform appearance, the layered geology beneath one's feet is complex. In some sections lava flows alternate with sedimentary deposits; in others the basalt is so deep and heavy it is believed to be pushing the valley floor down with it. All these volcanic permutations are born of one steady process: the slow crawl of the earth's crust across the Yellowstone Hot Spot, still seething impatiently beneath the Yellowstone Caldera. Place names like Bear Trap, Hell's Half Acre, Craters of the Moon, and no less than two City of Rocks lend relief—and fair warning—to the landscape. Except for the stray tuft of fescue, wild buckwheat, or stunted littleleaf mock orange prying its way into a crevice, the land looks raw, fresh as any new flow on Hawaii's Kilauea. Older deeper soils roll away under the standard sea of sagebrush.

The Snake River and the aquifers below this country provide ample resources for agriculture, and farms paint a green stroke across an otherwise burnt brown land. From Ontario, Nampa, and Boise, pockets of cultivation alternate with somber basalt cliffs and black lava flows, even sand dunes, all the way to Hagerman and Idaho Falls. Northeast of the Snake River Plain, the northern fringe

Rugged terraces of
Oregon's Imnaha Canyon.

of the basin and range pattern reappears, a complex jumble of volcanic and sedimentary mountains. Steppe vegetation cloaks each valley, lapping up and over low gaps in the Continental Divide at Idaho's ragged border with southwest Montana. Here at the headwaters of the Missouri River, the grassy sagebrush steppe makes a subtle transition into the prairie of the northwestern Great Plains, retreating steadily eastward.

South-central Oregon is vast and empty, undulating lava plains etched by seasonal watercourses looking for a way out, but there is none. Shallow drainages converge in saline playas or high desert wetlands. Thick layered basalt flows still crease every horizon here, but the tectonic roots of the basin and range are now evident. The wide escarpment of Steens Mountain leaps a mile above the saline flats of the Alvord Desert, a classic fault block model for all that lies south of here. The sheer isolation of Steens Mountain makes it a wild and unique place, even compared to its remote surroundings. Ancient glaciers carved U-shaped cirques into its basalt cap, now lush with meadows harboring several endemic plants found nowhere else on earth.

Back near Bend, Oregon, an imperceptible divide leaves the Deschutes watershed behind, and brings you to the Klamath River basin. The inland territory

The Alvord Desert
below Steens
Mountain.

south past the volcanic rubble of the Newberry shield supports open forests of rust-plated ponderosa pine rooted deep into the volcanic ash and pumice, dominant features all the way to the California border 150 miles beyond. At the source of the Klamath, wide shallow basins contain Oregon's largest lakes and jig-sawed wetlands. Though many have been drained and diverted for agriculture, these wetlands are a vital stop for millions of birds traversing the Pacific Flyway. Adjacent hills are a blend of steppe vegetation and forest, slopes gray with sage and shadowy junipers.

Landmarks like Devil's Garden and Ya Whee Rim carry on the volcanic theme of the Cascades, but the range is lower here. Low enough, in fact, that the Klamath River manages a Houdini-style escape, taking an unlikely westward turn under the still waters of Lake Ewauna and slipping out of the steppe and down its remote canyon. Past the temporary restraint of several reservoirs and pressed between the folds of the Siskiyou and Klamath Mountains, whitewater rapids make a steady descent, eventually joined by the Trinity River, making the Klamath California's second largest, and like the Fraser and the Columbia, the only other river to escape the interior steppes for the Pacific.

Otherwise, the western boundary of the Great Basin is essentially an unbroken continuation of the Cascade Crest to the north. Nearing the Oregon-California line, the Coast and Cascade ranges unite in a jumble of oak-studded hills, rumpled ridges still draped in conifers, verging finally on the misty summits of the Siskiyou and Klamath mountains. Rugged and remote, the Klamath reach heights over 7000 feet within 30 miles of the shore, forming an effective rain shadow. Their eastern flanks jostle with the southern terminus of the Cascades, which rise and fall as they soften into the Modoc Plateau to the east.

The volcanic Cascades of northern California are punctuated by ethereal Mount Shasta floating above the valley haze, and to the southeast Lassen Peak, but otherwise they seem to merge into the Sierra Nevada with scarcely a difference to the eye. However, their origins are distinct, the Sierras resulting not from recent volcanism but from the tectonic rise and fall of fault block systems, resulting in dramatic eastern escarpments that mark the edge of the Basin and Range Geologic Province.

The Sierras continue uninterrupted some 400 miles south, their abrupt eastern slope plunging into the vast basin and range topography beyond. The height of the Sierras rises continually from 7000-foot peaks and ridges in the north to the sky-shredding 14,000-foot summits of Russell, Muir, and Whitney, then tapering again toward 3793-foot Tehachapi Pass.

The world drops away into a network of deep valleys east of the Sierras. Alkali-crusted valley floors shimmer under parched blue skies. Where vegetation does exist, black greasewood, fourwing saltbush, and other shrubs soften the blinding alkali glare. This is the salt flat equivalent of sage scrub, occupying the hostile, more saline soils that sage will not tolerate. This habitat is typical of remnant and seasonal lakebeds throughout the Great Basin. In favorable locations, *Krascheninnikovia lanata* (nearly identical to its Mongolian counterpart, *K. ceratoides*) brightens the mix with its silvery leaves and seeds.

Fed by rain and snow from a wetter time, deep waters once filled these wide valleys. Some estimate that at its wettest, thousands of square miles—as much as half of the Great Basin area—may have been covered in water. A drying climate doomed these waters to a slow evaporative death, or at least a decline. The remnants of these persist in the shallow lakes and rich wetlands of Upper Klamath, Goose Lake, and Summer Lake. More often though, meager water sources were far outpaced by the evaporative duo of heat and wind. The pale ghosts of Pleistocene lakes Chewaucan, Modoc, Lahontan, and others now glaze the windy flats in gritty white minerals.

Desert lakes are a recurring feature, especially in the western Great Basin where Sierra snowmelt runs east into austere valleys. The Carson, Walker, and Truckee rivers are among those that begin as Sierra snow and meet a similar fate in captive lakes and playas, providing the unlikely spectacle of American white pelicans circling down toward their mirrored reflections. Even the indigo waters of Lake Tahoe are part of this closed system, spilling down the Truckee River to the brackish waters of Pyramid Lake.

The Ruby Mountains are a prominent feature in northeast Nevada, cresting above 11,300 feet, their mid and upper slopes well forested with aspen, limber pine, and whitebark pine. Winter snows can be deep, providing a steady runoff each spring, and summer thunderstorms help maintain perennial streams. Well-watered valleys below are a virtual oasis compared with much of the region, threading through the rangeland northwest of the Rubies to create prime ranchland around the towns of Lemoille and Spring Creek.

A few rivers are born and die here in the heart of the Great Basin. The Humboldt is the largest of these, its many branches trickling down from the Ruby range and weaving between disjointed mountains to water the towns of Elko, Carlin, and Winnemucca. Its 300-mile course, still followed by Interstate 80, was the favored—if not only—route of pioneers crossing the desert for the promised land of California. The river meanders west in broad curves between the ends of parallel mountain ranges as if it has somewhere to go. In a final flourish, it greens a lush grid of alfalfa fields at the oasis town of Lovelock, then disperses through the wetlands of the Humboldt Salt Marsh to end its journey in a sun-bleached playa.

On favored summits high above the sagebrush steppe, life has persisted relatively unchanged for thousands of years, a testament to the entrenched patterns

Near St. George, Utah, the Mojave Desert, the sagelands of the Great Basin, and the Colorado Plateau converge.

that strike a balance over time. In the White Mountains live the oldest known trees in the world, the Great Basin bristlecone pine, *Pinus longaeva*. Persisting through cruel weather and cyclical swings of drought, some Great Basin bristlecones have seen more than 5000 summers and winters come and go. Even individual needles may live for three or four decades before dropping. They live in scattered populations, always at cool high elevations, across central Nevada and Utah. The similar Rocky Mountain bristlecone (*P. aristata*) is distinguished by white resin dots on the needles and ranges mainly across central Colorado and northern New Mexico, also at high dry elevations.

The southern boundary of the Great Basin steppe is at best an approximation, as the basin and range topography changes little overall from north to south, but the higher Sierras here create a most dramatic rain shadow, placing this region in the milder, drier realm of the Mojave Desert. An irregular eastward line can be drawn from the Sierras to the Pine Valley Mountains of southwest Utah, wherever the desert-loving creosote bush (*Larrea tridentata*) community yields to more cold-tolerant species like blackbrush (*Coleogyne ramosissima*), saltbush (*Atriplex* spp.), and galleta grass (*Pleuraphis jamesii*).

The eastern edge of the Great Basin steppe is easily delineated by the dramatic north-south fault block of the Wasatch Range, carving Utah neatly down the middle. At its highest, this western branch of the Rocky Mountain system reaches heights over 11,000 feet, yet another rain shadow lying to its east. The Wasatch are crossroads of sorts, influenced by the Northern and Southern

Rockies, the Great Basin, and the Southwest deserts, all intergrading with the sagebrush communities.

The northern Wasatch Mountains join the splintered scraps of the basin and range, mostly forested here, with lower slopes cloaked in grassland, sage, and curl-leaf mountain mahogany (*Cercocarpus ledifolius*). Irrigated patchwork farms green the valleys. The Bear River loops quietly around the northern tip of the Wasatch, destined for the Great Salt Lake. It is this unassuming gap that gave passage to the legendary Oregon Trail, and to the east, the sprawling Wyoming Basin opens wide, the quintessential landscape of sagebrush steppe rolling to all horizons. Here even the Continental Divide loses direction, splitting to enclose a great landlocked bowl. Sage and prairie swell imperceptibly over the divide and wash out across the Great Plains, engulfing Laramie, Casper, and Gillette in a sea of grass.

The Wasatch stretch south for nearly 350 miles, rising abruptly in places, though not in a continuous vault like the Sierras. Hayfields fill broad valleys, and sagebrush spills through piñon and juniper passes into the Colorado Plateau to the east. Their southern regions host an interesting chaparral community of diverse elements, including *Acer grandidentatum*, *Quercus gambelii*, and *Ceanothus velutinus*.

Hydrology along the western Wasatch Front is in many ways the mirror image of what happens in the east Sierras. Melting snows in the Wasatch make a speedy descent down the west slopes, the Sevier River pooling into land-locked Sevier Lake; the Provo draining into Utah Lake; the Weber and Bear rivers into the Great Salt Lake. As expected, these lakes have their own unique history.

The Bear River used to flow north into the Snake River until lava flows diverted it into ancient Lake Bonneville. This increased flow helped to fill the entire basin 1000 feet deep. The largest of all the prehistoric pluvial lakes, it grew to about the size of Lake Michigan! Its bathtub rings of ancient terraced shorelines tell a story of fluctuating water levels over several million years. Even then this lake usually had no outlet to the sea, occasionally topping the gravelly soils at its northern extremities and into the Snake River. That changed drastically when its waters reached a critical level and breached the alluvial dams that retained its northern shore. All gave way in a spectacular flood similar to those of lakes Columbia and Missoula, though occurring several thousand years earlier. Its new outlet was quickly deepened as water fled north through Red Rock Pass by the cubic mile, carving through hundreds of feet of bedrock and deepening the chasm of the Snake River and the Columbia.

As the climate became drier, evaporation lowered the waters well below its outlet until a tenuous equilibrium was reached. The Bear River's course was set, and it still flows into the diminished waters of the Great Salt Lake, its flow just enough to maintain the lake near its present level, but it remains the largest river in North America to be denied its exit to the sea. Increasingly salty waters were simmered down to their bare elements, leaving behind the blinding Bonneville Salt Flats, a bleached bone of a desert visible from space. A dust storm in this playa could pass for a typical Sunday afternoon on Venus, air and earth and sky smoky-white, only gravity to keep you oriented feet down.

The Great Basin is a stronghold of space, and one of the best places to contemplate life on the surface of a planet in slow but steady flux. The quintessential experience is achieved by driving southwest from the urban Wasatch corridor, continuing west on Highway 50 from the sleepy town of Delta, Utah. For much of its 400-mile length, this "Loneliest Road in America" spends its time climbing more than a dozen mountain passes or sweeping low into lunar valleys that dip to the vanishing point before tilting imperceptibly toward the next mountain horizon, the scent of sun-warmed sage and juniper on the air. The trappings of civilization seem especially out of place here—a car, a phone, an apple, a pair of shoes—all foreign objects not of this stony world. I stopped at no particular place once and walked a slow half-mile and back down the middle of the empty road (because I could), an immersion experience into the silent language of a primal world wheeling steadily on its way.

WINDING THROUGH THE COLORADO PLATEAU

The Wasatch Mountains are the backbone of Utah, their northern peaks gleaming above Provo, Salt Lake City, and Ogden. Their deep snows support legendary skiing, and send precious water to the cities below. East of the Wasatch Divide, they make a relatively gradual descent, hills dotted with piñon pine and Utah juniper, slopes and valleys splotched silver and lime with the predictable cover of sage, ephedra, and rabbitbrush.

But 15 miles east of the deep-powder ski towns of Alta and Park City, an anomaly begins. The mountains of the Uinta Range claw their way out of the forest to barren heights over 13,000 feet, but they defy the north-south pattern of nearly all North American mountain ranges. They run east-west, effectively dividing the Wyoming Basin to the north from the Uinta Basin and the Colorado Plateau to the south. They have seen ice ages come and go, leaving them gouged with U-shaped valleys and dotted with hundreds of lakes. Lower slopes and plateaus are a mix of sage-covered mesas edged with badlands, and increasingly pockmarked with energy platforms in search of oil and gas. Below them the basin opens wide with hayfields.

Deep sedimentary deposits hold more than fossil fuels. Some of the richest deposits of dinosaur bones have been found in the ravines that slice through the plateaus of this region. The canyons of the Green and Yampa rivers converge

ABOVE An idyllic road winds through Cathedral Valley in Capitol Reef National Park.

RIGHT Tough but beautiful plants thrive in southwest Wyoming's hardscrabble expanses.

The Green River slips quietly through Colorado's Brown's Park.

near the town of Vernal, exposing a treasure trove of fossils from primal tropical worlds that seem the polar opposite of current high desert conditions.

The Green River makes a wide turn, taking the long way around the east end of the Uinta Range. By the time it reaches Vernal, it has traveled nearly half its length, spilling down out of the snowfields and lakes of Wyoming's Wind River Mountains and across the weathered surface of the Wyoming Basin. The expansive Green River drainage is high sagebrush steppe at its finest, essentially uninterrupted for 250 miles from Bondurant, Wyoming, to Hayden, Colorado, except for the occasional ravines or badlands carved into the dry hills. To distinguish this basin from those that adjoin it seems trivial to most people passing through, as the endless sage covers most of this windy country, no matter the watershed. Winters here are frigid and summers short, so other than grazing, most agriculture is doomed, leaving vast stretches unspoiled.

Along the Green River's course, canyon-style topography is interspersed with grassy floodplains until it nears the Utah border and Flaming Gorge. The Green plunges through layer upon layer of warped and uplifted sedimentary rock in a series of canyons, through the Gates of Lodore and Split Mountain, around the shallow meanders of Horseshoe Bend, then deepening again into Desolation Canyon. The river drops only 700 feet from Split Mountain south to its exit from Desolation, but in places the canyon rims of the Tavaputs Plateau tower 5000

The San Rafael Swell
rises like a storm-
driven sea of stone.

feet above the river, the entire plateau having tilted up beneath the established meanders of the river, grinding its course steadily deeper into ancient rainbow-layered stone. The south face of the Tavaputs and adjoining plateaus form the Book Cliffs, a 200-mile-long S-shaped front that reaches from the Wasatch to Grand Junction, Colorado. Similar topography and transitions can be found in other steppe regions, like the fissured western edge of the Gobi Desert in Mongolia, and the Taklamakan Desert where it rises up to meet the south edge of China's Tien Shan mountains.

The Colorado Plateau is a geologist's dream, water having opened the pages of the earth for all to read. There are plentiful perennial streams and rivers that wind their way between blocky plateaus to the Colorado River. In the west, from the Wasatch Mountains and the Utah plateaus come the rivers Price, San Rafael, Fremont, Dirty Devil, Escalante, and Paria. These and their tributaries pass through hundreds of miles of dramatic gorges and slot canyons, slicing against logic through the tilted beds of rock at Capitol Reef and the San Rafael Swell.

Anticlines, synclines, and monoclines have warped the earth, thrusting fins of rock in waves that seem to still be in motion, the rivers relentless in negotiating the rising tide of stone.

From the east, the rivers Colorado, Gunnison, Dolores, and San Juan melt out of the Southern Rockies through miles of spruce and aspen, into increasingly arid country. Irrigated valleys of Colorado's West Slope are lush with hayfields, peach and apple and pear orchards, beans and sunflowers and sweet corn, set against ruddy piñon-covered hills and adobe badlands. The waters often drop out of plain view into deep canyons.

The Colorado River Valley opens and closes in a series of canyons and wide valleys, finally spilling out of its red and brown walls at Palisade to water the towns and orchards of the Grand Valley. The Gunnison slices the Black Canyon in a sheer head-spinning half-mile drop below its gently sweeping sage and piñon rim. Joined by the Uncompahgre River, it continues on, out of view for much of its length along the base of the Uncompahgre Plateau to join the Colorado at Grand Junction.

The Dolores River makes a long and winding descent out of the San Juan high country. In a classic example of a meandering stream carving into the land as the earth rose beneath it, it makes tight turns deep between red stone walls. At the town of Bedrock, it bursts out of its canyon walls and slips across the flat Paradox Valley floor, the paradox being that it crosses perpendicular to the valley and disappears back into its deep canyon walls on the opposite side, following a channel that was set in stone long before the fault block floor dropped from under it.

The San Juan River also leaves the high country of Colorado to wind its way in and out of New Mexico and into Utah, passing within a mile of the Four Corners. Confined to a tortuous route between layered walls of red and brown shale and limestone, it snakes through remote country more than 1000 feet below the rim. It joins the Colorado River with little fanfare under the impounded waters of Lake Powell.

Higher elevations throughout are a puzzle of piñon-juniper woodland interspersed with sage flats and sparse grasslands. Life here exists on a smaller scale as well. Biological soil crusts, a shared component of arid climates around the world, are especially well developed in the sage and piñon-juniper habitats. These complex communities of cyanobacteria, lichen, algae, moss, and fungi are essential to the long-term health of native plant communities, stabilizing soils and enabling penetration and retention of scarce precipitation with their corrugated surfaces. They are easily disturbed by wayward feet or wheels, taking many years to fully recover, and once destroyed, the door is open to erosion and invasive species.

On the far horizons, widely spaced volcanic mountains loom in isolation. The Abajo range, the La Sals, the Henrys, Navajo Mountain—these are the sky

Dry washes typify this land of plateaus and canyons.

islands of the Colorado Plateau, their slopes dense with pine and aspen forests, and summits glinting with melting snowfields well into the summer. They provide relief to the eye but also harbor a wealth of wildlife and wildflowers not adapted to the drier conditions that surround them at lower elevations. These are the sources of perennial streams that thread through an otherwise dry landscape.

To the uninitiated, the broad plateaus between these mountains can have a monotony of sorts, nearly level or barely rolling, dense with sage scrub and patchy piñon and juniper in shades of black-green and gray. Many plateaus are table flat, others list like sinking ships. At a turn though, the road may warp downward into an unforeseen chasm. This is the most colorful of the intermountain steppe regions. Iron-rich sandstones dominate the horizons, standing upright in spectacular rust-toned monuments, or sliced abruptly away into twisting canyon mazes. Against this backdrop, the color of every living thing is enhanced, from silver-green roundleaf buffaloberry to the acid-green of ephedra. Scarlet penstemons ignite the side of an arroyo, the dry blue expanse of space overhead. Thousand-foot walls of rich red Wingate sandstone hold up the sky, capped with buff domes of Navajo sandstone and elfin forests of piñon pine and Utah juniper. Soil depths vary dramatically, deep windblown red sand providing footing for sagebrush, and jagged outcrops swept clean, with ricegrass, prince's plume, or globemallow anchored in hospitable crevices.

Below the Glen Canyon Dam, the Colorado River enters its own unique realm of ever-deepening canyons, the ultimate being the Grand Canyon itself. Its great depth is due not to the sudden plunge of the river, but to the bulging plateaus that rose beneath it over millions of years. This now-familiar theme is expressed time and again in the regional landscapes. Theories abound on the exact mechanisms that created this spectacle. However, it is generally agreed that the adjacent Uinkaret, Kanab, Shivwits, Coconino, Hualapai, and Kaibab plateaus were all part of the Kaibab uplift that set the river carving like a bandsaw into the ancient layered rocks. The mile-deep gorge basks in desert heat, and as the river slips out of the canyon beyond the Grand Wash Cliffs, the Mojave Desert simmers with creosote bush and Joshua trees. But for the plants, the rugged landscape of the Lower Colorado River beyond could pass for Pakistan's Indus River as it carves its way below the steppe between the Hindu Kush and Himalaya near the town of Chilas.

These plateaus north and south of the canyon rise to between 5000 and 9000 feet, with a full complement of habitats, from sagebrush and grasslands through aspen and ponderosa forest. This forms the southwestern extent of the sagebrush steppe, though a few mountain ranges beyond here still have pockets of steppe-like vegetation, even reaching the upper Verde River watershed in Arizona's Chino Valley.

The surrounding tablelands extend east and south into northern Arizona and New Mexico, taking in the drainage of the Little Colorado River. Upper portions of the river are perennial, fed by snows and springs in the White Mountains and the gradual north tilt of the pine-forested Mogollon Rim, but below St. Johns the river is ephemeral. Its dry wash continues toward the Grand Canyon through the Painted Desert and past lava flows and petrified forests, dropping between narrow brown walls to join the Colorado River. Summer thunderstorms may send torrents of chocolate water hurtling over cliffs and sandbars, but for most of its life it lies dry and silent. This area is desert, but true to the pattern, higher elevations nearby transition into sage, greasewood, or saltbush flats, grassy savannas, and piñon-juniper woodlands.

South and east of here the Continental Divide scrawls a ragged line over the

Reflective
pools
remain after
summer
storms.

tablelands and canyons of western New Mexico. As happens in Wyoming and Idaho to the north, the steppe spills past the divide unnoticed, transitioning into the Chihuahuan desert grasslands to the south. Key plants give away the difference—the trunk-forming soaptree yuccas, and honey mesquite, for example—but these are true grasslands, and it could be argued that these southerly regions are but a milder version of steppe, extending well into the highlands of northern Mexico. Similarly, south of Arizona's White Mountains the pine-oak woodlands and savannas drop into the Madrean Basin grasslands. High elevations bring you to similar life zones as would be found farther north: shrublands, piñon-juniper woodlands, ponderosa pine forests. Perhaps the greatest distinction here from true steppe is the lack of sustained winter cold, allowing a dramatic increase in plant diversity. Additional elements of the Madrean flora include many oaks, manzanitas, yuccas, cacti, and agaves.

The steppe landscapes of the Colorado Plateau are woven into many of the world's legendary wonders—Arches National Park, Bryce Canyon, Zion, Canyonlands, Capitol Reef, Grand Canyon—and countless other wonders rich in solitude and off the beaten path. To take in any of its thousands of canyons and plateaus is to experience the intermountain steppe at its colorful best.

Plant-People Connection

The long history of human occupation in North America has already filled many literary volumes. It's difficult to synthesize the salient points of reference in a relevant way, but it is believed that people migrated to North America some 30,000 years ago, crossing the Bering Land Bridge from Asia when the waters of the world were bound up in great ice sheets, leaving the oceans some 300 feet lower than they are today. What happened beyond that time is a matter of research and hypothesis. Was the Columbia Basin one of their first homes? How did ancient people view this calamitous landscape? How did they survive and adapt to the steppe and its challenges? Did they find better habitation near the coast or were the dry interior valleys more hospitable? Did they witness the retreat of the ice sheets, the vanishing land bridge, or the end-of-the-world floods of Lake Missoula? Many scenarios are possible, but whatever the precise timing of these events or the course of expansion of the earliest Americans, they found plentiful resources in the Northwest. The current landscapes would likely have been familiar to them, the mountain ranges firmly in place, rivers still careening to the sea.

In the Pacific Northwest, raw materials were abundant for the earliest inhabitants. Rivers churned with salmon by the millions, and favored fishing sites were hubs of habitation, migration, and trade. Many Native Americans still live in the region, their tribal identities familiar now as the names of towns and landmarks:

Petroglyphs on the Colorado Plateau: old enough to be mysterious; young enough to show the presence of the horse.

Yakima, Palouse, Nez Perce, Umatilla, Spokane, and dozens more throughout the region. Their history cannot be summarized fairly in a few lines, but it echoes the lamentable fate of most indigenous peoples: introduction to Europeans, spread of disease, loss of territory, violent conflict, marginalization, vital habitats plowed or flooded, difficult integration into a changed world. This disturbing reality is forever etched in the human landscape.

Even in the harsh expanses of the Great Basin, resources were adequate to sustain early inhabitants. High elevations held water, game, and respite from summer's valley heat. Low elevations were an escape from deep snow and cold. Still lightly populated, its exportable resources underpin most communities today: minerals, metals, beef. Where these are scarce, life is a gamble, literally, and casinos allow the masses to support unlikely cities like Las Vegas.

The Colorado Plateau has been home to Native Americans for thousands of years, their histories etched or painted on hidden canyon walls, stone foundations of dwellings still stacked as they left them, tumbled walls holding caches of blue juniper berries. Tree rings of timbers found in such dwellings help date their culture and also reveal dramatic climate cycles of abundance and drought, severe enough to close the chapter on Anasazi history. This has always been and continues to be a climate in flux. We often see only the immediate storm or drought or fire, but plant communities can shift over long periods of time, changing the landscape as forests and shrublands climb or descend in elevation.

Still, the earliest inhabitants came to know this land and its rhythms for millennia. Around 1700, the introduction of the horse—a perfect match to the interior steppes—expanded their trade routes from the Pacific to the Great Plains, and their own cultures changed as a result. In the north they met the horse some years before most had met actual Europeans, a result of native trading networks that effectively linked them to early Spanish colonization hundreds of miles away.

Colonial motives spurred most early ventures into the arid west. Though native people knew well how to live in this land, most of the interior was an utter mystery to Europeans. Was there a water passage to the Pacific? Lost civilizations? Rich soil? Gold? By the early 1800s, 300 years had already passed since the Spanish—compelled by religious zeal and a taste for gold—had ransacked Mexico City and began their steady expansion. Missions in Arizona and New Mexico were already 200 years old. In all this time, fitful excursions by Russian, French, and Spanish vessels had expanded knowledge of the Pacific Coast, and fur traders had become a fixture. George Vancouver's expedition of 1791–95 explored the shores in more detail, even venturing about 100 miles up the Columbia. Accompanied by Archibald Menzies, they collected new plant and animal specimens along the way, but this still left most of the interior to the imagination.

Thomas Jefferson, having completed the Louisiana Purchase, was committed to further exploration of this territory, and the obvious next step: to lay claim to the unknown lands all the way to the Pacific. A potential, though unrealized, water route across the continent was a primary motivation. With Jefferson's backing, the Corps of Discovery departed in 1804 for the unknown, led by Meriwether Lewis and William Clark. Their travels and discoveries are well documented and fascinating to read. Unknown botanical specimens collected along the way were but a small sampling of the diversity that marks this enormous country.

Not all early explorers were bent on conquest, not directly at least. Pressed plant specimens were gathered by the thousands, and though some were lost to unfortunate calamity, many were shipped far and wide to university herbaria and taxonomists in East Coast cities, or on to England and Europe, where scientific names were conjured at an unprecedented clip. For botanists and gardeners, the list of botanical explorers will ring familiar, their legacy preserved in hundreds of epithets, among them Thomas Nuttall (*Calochortus nuttallii*), Nathaniel J. Wyeth (*Wyethia amplexicaulis*), and John C. Fremont (*Physaria fremontii*), to name a few.

While most early explorers were collecting pressed specimens for research and identification, all in the name of science (and notoriety), a few also had an eye for horticulture. David Douglas was sent to the Northwest in 1824 by the Royal Horticultural Society in England with the express purpose of collecting new and unknown specimens, and adding botanical diversity to their gardens. He was an intrepid traveler, eventually meeting an untimely death in Hawaii at the age of 39, but not before his multiple expeditions had identified hundreds of new plants and contributed well over 200 new species to the known world of horticulture, among them the namesake Douglas fir.

Despite this early interest in all things novel and wild, commerce and survival hinged on meeting human needs. Native plants and landscapes painted the irresistible images that drew immigrants to the wilderness of the Northwest, but once settled, most things wild were grazed, chopped, or plowed under for production of necessities and profits. The richest prairies went first, followed by deep soils of sagebrush "deserts" reclaimed in the name of progress. The conversion of sun and soil and rain into hay and apples and cherries made towns, cities, and new lives possible.

Native plants usually stood in the way of things more profitable, but not all went unappreciated. Savvy gardeners in Europe had reputations to maintain, and having the latest and greatest from the wilds of America was a feather in the cap. American plants found a ready audience waiting with open arms and furrows for the likes of *Lupinus polyphyllus*, *Cornus nuttallii*, *Abies procera*, and on and on. Most of these came from wetter mountain or maritime regions, a better cultural match for the humid gardens of Europe. Within the steppes it was primarily damp meadows that yielded plants of interest—*Dodecatheon*, *Iris*—and many of these (or their relatives) were already known from farther east. Bulbs were of interest—*Camassia*, *Allium*, *Calochortus*—but those from dry habitats were difficult in cultivation, rotting away in soggy British gardens. Showy perennials like *Penstemon* and *Eriogonum* in their many forms were short-lived as well. Early rock gardens designed for alpine plants were better habitats for them, but this was for the specialist, not the average garden plot. Thus most of the unique plants of the true steppe went unnoticed, or at least uncultivated, until more recently.

The movement of plants was not a one-way process. As settlement increased, introduced plants from similar steppe climates in eastern Europe and Asia tell a cautionary tale, finding fertile ground and conditions just like home—cheatgrass (*Bromus tectorum*) being the infamous overachiever of the bunch. Introduced both accidentally and, in at least one instance, deliberately, it quickly escaped. To say it spread like wildfire is not an exaggeration; its highly flammable remains changed entire ecosystems by incinerating shrubs with poor resilience and crowding out native bunchgrasses. It now dominates millions of acres, especially lands that were degraded by disturbance or overgrazing, leaving many native plants at risk. It shares the blame with a suite of other poorly behaved invaders—*Cirsium*, *Hypericum*, a vicious handful of *Centaurea* species, and more—all perfectly at home in this steppe-sister world.

Another small but vibrant endemic, *Penstemon acaulis*, Uinta Basin.

RIGHT *Eriogonum bicolor*, endemic to the Utah canyonlands and plateaus.

Habitats and Plants

The vegetation of the intermountain steppe follows predictable patterns that can be found in all steppe regions of the world. Corresponding latitudes of Central Asia, South Africa, and Patagonia could present a similar vision. Plunked down in the middle of any steppe region, you might be hard-pressed to declare with assurance exactly where on the globe you had landed. Is this the salt-crusted shore of Nevada's Pyramid Lake? Or maybe Tajikistan's Karakul Lake? Patagonia's Lago Posadas or Cardiel? Or Quaggaskloof Lake in South Africa? Geologic origins may be different, but plants the world over use similar adaptive strategies to survive the similarly arid, sunny, windy conditions. This results in predictable vegetation patterns, and there are even shared genera of plants that occupy parallel niches in these far-flung regions, a relic of the congealed supercontinents of pre-history. A close look at the native vegetation can help solve the riddle, but these places share the experience of openness, expansive skies, and bright, dry, sunlit vistas with naked geology lying sprawled out to the horizon.

Elevational differences, local variations in precipitation, and ravines or riparian corridors create rich plant communities, but these are anomalies within the vast "space between." This range of habitats produces a wide diversity of plants. Many plants that are found in the sagebrush steppe also overlap into adjacent areas. Genera like *Artemisia*, *Eriogonum*, and *Ericameria* range widely from California and Mexico far into Canada and the Great Plains, sometimes as the same species, sometimes as unique species, varieties, or subspecies. Delineating precise boundaries is impossible.

Some plant families contribute a great diversity of species to the mix. Others —Asteraceae, Brassicaceae, Amaranthaceae, Cyperaceae, Poaceae—may dominate the landscape with vast monocultures that can extend for miles, such as saltbush or alkali sacaton. Several families and genera are shared with the other steppe regions, and related plants sometimes run amok when introduced to their sister climates. Here the quintessential western plant—tumbleweed (*Salsola tragus*)—spins across a landscape just as hospitable as its native Russia.

Other families or genera may be equally diverse, but not all are abundant. Many of their members lay dispersed in isolated endemic populations—those found nowhere else in the world. This includes *Astragalus*, *Erigeron*, and *Penstemon*, to name a few. Specific rock strata, chemical composition of soils, limited range of pollinators, physical or genetic isolation over time, and changing climate can all contribute to the rarity of endemic plants. Human disturbance threatens them even further. These habitats account for numerous rare and endangered species that call for special protection and study. The Colorado Plateau has the highest rate of endemism in the Intermountain West.

Blue-stemmed *Ephedra nevadensis* is abundant in the Great Basin.

SALINE BASINS

Lowest elevations, primarily where water has no outlet, are typically saline basins. This is usually the result of ancient lakes receding where incoming water cannot keep pace with rates of evaporation, leaving saline deposits behind. These are sometimes devoid of all vegetation due to extreme salinity, but their fringes may support plant communities dominated by salt-tolerant shrubs, such as saltbush (*Atriplex confertifolia*, *Grayia spinosa*, and others) and black greasewood (*Sarcobatus vermiculatus*). These communities are typically low in diversity, and sometimes exist as near monocultures. A handful of grasses (including *Pleuraphis jamesii* and *Sporobolus airoides*), sedges, annuals, and perennials may be scattered within these habitats. The presence or absence of species is determined by a range of factors, including salinity, soil type, water table, rainfall regimes, and grazing pressures. Fringes and upstream basins where freshwater passes through usually have entirely different plant communities, not limited by the saline conditions. Where drainage is adequate, *Ephedra nevadensis* or *Krascheninnikovia lanata* may be abundant. In horticulture, few plants from these saline habitats have found a wide usefulness, though some have unrealized potential for reclamation under difficult conditions.

NON-SALINE LOWER TO MIDDLE ELEVATIONS

These habitats include alluvial fans at the base of mountains, sandy soils and dune fields, and low-elevation plateaus and valleys that are drained by watercourses or

Cleome serrulata is a favorite of native pollinators.

seasonal washes. At their lowest elevations (generally below 5000 feet) in the southern reaches of the steppe, *Larrea divaricata*, which dominates so much of the true Mojave Desert, loses its monotonous grip, replaced primarily by sagebrush and saltbush. Some plants from adjacent desert areas continue into the lower, drier steppe (sometimes referred to as desert steppe), including blackbrush (*Coleogyne ramosissima*) and even Joshua tree (*Yucca brevifolia*), found where sage and piñon-juniper transition into Mojave Desert shrubland; but these are primarily high desert plants found only in scattered habitats.

Many associated steppe grasses and wildflowers make an initial appearance here, and their ranges continue into the lower reaches of the sagebrush steppe. Lower elevations of the Colorado Plateau are especially diverse with their deeply cut canyons allowing a rich mix of sagebrush, desert, and riparian communities, depending on aspect and exposure. Indian ricegrass (*Achnatherum hymenoides*) helps stabilize sandy dunes and forms tawny tufts among eroding sandstone ledges and boulders, alongside pink and purple *Cleome serrulata* and the silver wands of sand sage (*Artemisia filifolia*).

Asteraceae, Amaranthaceae, and Fabaceae are especially well represented; commonly found genera include *Artemisia*, *Ephedra*, *Yucca*, *Eriogonum*, *Ericameria*, and *Opuntia*. A number of plants (several species of *Astragalus*, the colorful

Stanleya pinnata) are associated with soils that are high in the element selenium, which is absorbed into their tissues; this can be toxic to grazing livestock and wild animals if other forage is scarce.

SAGEBRUSH COMMUNITIES

Areas dominated by sagebrush vary with latitude, exposure, and soil conditions. The precise mix of plants also varies from the Columbia Basin to the Great Basin to the Colorado Plateau, but many qualities are shared throughout the region. In the southern part of its range sagebrush may occupy cooler high-elevation valleys or horizontal zones at favorable elevations where precipitation is slightly higher than surrounding valleys, but not yet enough to support continuous forest. Sage communities can be found as high as 10,000 feet in some areas. As one moves north through the region, these zones become more continuous, with sage dominating entire valleys and intermixing broadly with piñon-juniper forests and even filling openings in ponderosa or limber pine forests.

This is classic sagebrush steppe, rich with composites—all Asteraceae—including at least five species or varieties of *Artemisia* dominating the landscape for thousands of square miles. *Artemisia arbuscula* is more frequent in parts of the western Great Basin, increasingly mixed with or replaced by *A. tridentata*, and farther east, with *A. nova*. More colorful composites are abundant as well, and each takes the spotlight in its own season, including species of *Symphyotrichum*, *Erigeron*, and others. An assortment of bunchgrasses are usually part of the mix and may even be dominant, especially in areas that have not been overgrazed, or areas that receive a bit more summer precipitation. In favorable conditions, several geophytes thrive including several species of *Allium*, *Calochortus*, and *Camassia*. Penstemons are among the stars of these communities, a nectar-rich resource for insects and hummingbirds, and a colorful counterpoint to the gray-green landscape. Sagebrush steppe has contributed many valuable plants to the world of horticulture and is an indispensable resource for anyone gardening in the Intermountain West. Much work remains to be done with the native ornamentals from this zone; the surface has barely been scratched when it comes to selection or breeding of superior forms.

PIÑON-JUNIPER ZONE

This is the dominant forest mix throughout much of the region, so abundant that it claims the common moniker "P&J." Wherever elevations are slightly higher and precipitation is more than about 12 inches per year, these elfin forests often share territory with sagebrush, sometimes rooted among rocks and ridges less hospitable to sage, but equally at home on flat mesa tops and the lower flanks of mountains. These forests used to be more open, the grassy hills peppered with globes of dark green. This is still seen in some areas, resulting in the famed New Mexico landscapes immortalized in the paintings of Georgia O'Keeffe and others, but in much of the region fire suppression has resulted in crowded stands with very little grass or herbaceous cover and choked with seedling trees. Now vulnerable to drought, insects, and disease, fires can rip through the squat canopies and open the way for weedy invasives. Interestingly, in prolonged periods of drought, the piñon pines seem to suffer long before the junipers; whole hillsides of pine may go brown, leaving islands of olive-green juniper relatively unscathed. Still, the piñon-juniper zone dominates huge expanses of the West in varying stages of health, providing a wealth of habitats and resources to the wildlife and the people who call it home.

As expected, there are significant variations in the plant palette throughout the region. For example, in the Colorado Plateau this zone is dominated by *Pinus edulis* and *Juniperus osteosperma*, with *J. scopulorum* occupying drainages with more moisture. In the Great Basin *P. edulis* is replaced by *P. monophylla* as the dominant piñon pine, still mixing with *J. osteosperma*, yet as one moves west, this juniper is increasingly replaced by *J. occidentalis*. Move farther north, and *J. occidentalis* dominates completely, with neither piñon pine making an appearance. In portions of the Wasatch Mountains, these communities are often replaced by a chaparral zone featuring Gambel oak (*Quercus gambelii*), bigtooth maple (*Acer grandidentatum*), and a cadre of rugged shrubs, reflecting ties to the Madrean flora to the south and west. This includes a wealth of xerophytic members of Berberidaceae, Rosaceae, and Elaeagnaceae.

CANYONS OF THE COLORADO PLATEAU

Lowest elevations of the Colorado Plateau are true high desert, cold in winter, hot in summer, and dry always. Page, Arizona, near the southwest center of this region, averages little more than 7 inches of precipitation, its sandy plateaus thinly tufted with Indian ricegrass, ephedra, greasewood, and saltbush. Elevations below 5000 feet, especially the broad flanks of the southern plateaus, support communities of *Pleuraphis jamesii*, *Aristida purpurea*, and *Coleogyne ramosissima*, while the mesa tops at nearly 6000 feet are deep in sage, piñon, juniper, and the beautiful *Shepherdia rotundifolia*. Infrequent patches of the rare blue *Ceanothus greggii* var. *franklinii* bristle with blue or purple flowers in April, a desert adaptation of the famed ceanothus of California chaparral and gardens. Sandstone shelf rock and ledges have thin soils but deep fissures, allowing tough shrubs like *Cercocarpus ledifolius* to take root. Wet seeps in vertical sandstone alcoves harbor maidenhair fern and delicate columbines. Deep shaded canyon bottoms shelter redbud and boxelder. The transitions between desert and canyon and mountaintop create endless habitat opportunities and spark endemism.

Sheltered ravines abound with rare microclimates and endemic species.

MONTANE ZONE

At the highest elevations throughout the intermountain steppe, coniferous forests predominate, but their composition varies widely. Predictably, the high mountains along the western fringe are dominated by Cupressaceae associated with the Sierra Nevada and Cascade ranges. In the east, the Wasatch and Rocky Mountains share many common elements, though each has its own peculiarities. The summits of the vast basin and range region each stand in their own isolation, and forests that crown them are likely remnants from a moister era when forests were more continuous throughout the basin. As one might expect, their composition shifts from west to east and south to north. A wealth of plants from these higher elevations have made their way around the world, and an exhaustive list would include such noteworthy plants as the Colorado columbine (*Aquilegia caerulea*) and blue spruce (*Picea pungens*).

Contributions to Horticulture

The natural landscapes of the West are rich with microclimates. Sunny slopes, cool shaded ravines, shallow soils, baking rock walls, saturated swales, reflected heat, windy ridges—all these habitats have their counterparts in the conditions we create in our human communities. The north side of a house may stay damp and cool when compared with a rock terrace against the sunny south walls. Shallow soils near a paved lot may provide desert-like conditions, but the perpetual shade of a nearby highrise acts as a cool shaded canyon wall. The same diversity in nature provides us with useful and ornamental native plants for every niche of the garden.

Many of the plants of the intermountain steppe have gone underused or underappreciated. People moving into these regions usually brought their favorite plants with them, no matter how ill-suited they were to these semi-desert conditions. It has taken more than a century to bring native plants into the mainstream consciousness and the garden. When paired with suitable plants from other steppe regions of the world, amazingly diverse landscapes with an authentic nod to their steppe origins may be achieved. The resources required by a traditional garden may be many times those required by a garden that uses the best native and adapted plants, in the proper locations. This becomes an important consideration, especially when water resources are under ever-increasing pressure from homes, industry, and agriculture.

A list of desirable plants numbering in the hundreds would be appropriate here. Some have gained recognition and deserved popularity in regional gardens and beyond. Others have untapped potential but may be hindered by slow growth habits or challenges in propagation or availability. Many have strong family ties to their relatives in the steppes of Asia, Patagonia, and South Africa. All could have a place in the well-designed garden.

INTERMOUNTAIN NORTH AMERICAN STEPPE

Yucca baccata, Grand Canyon.

Tiny *Yucca nana* near Moab, Utah.

ASPARAGACEAE

True agaves only flirt with the southern fringes of the steppe, mingling with *Yucca brevifolia* as the Mojave Desert loses its grip. Beyond here, several yuccas bristle among rocks and canyons even as far as central Nevada but are completely absent from the Columbia Basin. South into Mexico the diversity multiplies. Despite having many subtropical species, *Yucca* does not make the leap across the vast tropical regions of South America and is nowhere to be seen in the steppes of Patagonia. The most robust species is *Y. baccata*, found among rocky outcrops and canyon rims; with leaves as much as 2 feet long, it can be grown as a large sculptural plant in gardens that are too cold for agaves. *Yucca harrimaniae* is naturally occurring in sandy soils on shelf rock in the colorful canyons of the Colorado Plateau; leaf rosettes can reach 18 inches across, with each leaf fringed in curly snow-white filaments. Ivory-white flower stalks rise 18 to 30 inches above the foliage in early summer. Plants slowly multiply to form colonies. The diminutive gem of the genus is *Y. nana*, often considered a variety of *Y. harrimaniae*. Slow growing and thriving on neglect, mature rosettes may reach only 4 to 5 inches in diameter. This yucca is a treasure to rock gardeners and those seeking a western touch in even the smallest xeric garden.

Grayia spinosa, pink-seeded form.

AMARANTHACEAE

This family does not make the best-sellers list—yet! Most of its members are easily overlooked, in part because their habitats are often among the harshest and least visited. This includes barren clay flats, adobe badlands, and alkali playas where little else grows. Still, they are a common element of most steppe regions, thriving where other plants fear to tread. Many tolerate or even prefer halophytic (salty) soils. Common names of greasewood, saltbush, and hopsage are used interchangeably between the hundreds of species that grow worldwide. Conspicuous genera here are *Atriplex* and *Grayia*, with dozens of species; *Sarcobatus*, with two species, now resides in Sarcobataceae.

Atriplex and *Grayia* have the greatest potential, not for their flowers but for their papery winged seeds, which form and persist for many weeks, often aging through pastels of yellow or rose-pink, and drying to tan. A handful have been used in horticulture, but they remain a virtually untapped resource with great potential for further selection. Among them are *A. canescens*, with papery winged seeds in shades of yellow to pink, and *G. spinosa*, with some forms exhibiting seeds that are a car-stopping hot pink.

ASTERACEAE

The aster or composite family is impossible to sum up, brimming with a confusing array of daisy-type flowers, often in shades of yellow. So many, in fact, that they are often lumped together under the acronym DYC—damn yellow composites! But not all are yellow and not all resemble daisies. All the sagebrush belong here, along with rabbitbrush, sunflowers, goldenrod, coneflowers, zinnias,

Artemisia filifolia with common sunflowers.

vernonias, and hundreds of other plants both familiar and obscure. The steppes are rich with composites worthy of cultivation, and to highlight just a handful does not do them justice.

Artemisia tridentata (big western sagebrush) is the most widespread of the genus, rooting deep into the rich sandy loam of mid-elevation western valleys. It is a fragrant addition to a western garden, lending authenticity and softening transitional areas between garden and wildlands. Plants suffer with too much water. Seedlings often appear, allowing the gardener to replace poor specimens with vigorous new ones. *Artemisia filifolia* (sand sage) is perhaps the most graceful shrubby sagebrush. It thrives in sand but is amenable to cultivation, with feathery wands of silver forming a soft dome in the garden. Of all the sagebrush, it responds best to shaping; a hard prune each spring maintains a uniform habit.

Many "daisies" are commonly grown, but the xeric natives of the intermountain region remain underused. Among them is the stout little *Tetraneuris acaulis*, with its tight mounds of green or silvery leaves and cheerful yellow daisies perched on sturdy 4- to 6-inch stems for much of the summer. Senecios, often lacking ray petals, are common throughout the West, and some Patagonian species are nearly identical. *Townsendia* species range from simple rosettes of leaves to silvery cushions, all crowned with daisy-type flowers, and well suited

LEFT *Wyethia amplexicaulis* and alliums.

BELOW *Scabrethia scabra* gilds the Brown's Park formation in Colorado.

Mahonia fremontii in spring.

to rock garden conditions. Commonly known as mule's ears, *Wyethia* shows off with large sunflowery blooms above robust clumping plants; gold-flowered *W. amplexicaulis* is a common companion to sagebrush, alliums, and lupines, and farther northwest the snow-white *W. helianthoides* brightens entire meadows. In drier locations, closely related *Scabrethia scabra* shines, whatever its backdrop. These are only a sample of the western Asteraceae worthy of cultivation.

BERBERIDACEAE

The dryland barberries (*Mahonia*) of the West have many qualities in common with their evergreen Asian counterparts, so well known in cultivation. Several species are found, all becoming large shrubs and all bearing protection, not as thorny stems but as well-armed leaves, rigid and blue-green with prickly teeth set along their margins. This only adds to their beauty, though perhaps leaves them less amicable to work with. On the Colorado Plateau *M. fremontii* sports especially stunning mature foliage of glaucous blue with subtle venation and stout prickles, like a holly leaf. New growth is infused with rose and wine shades as it emerges, made all the more striking when the lemon-gold flowers appear in spring. Flowers are followed by attractive red berries that persist for many weeks. This shrub can become a gnarled and beautiful specimen in a xeric garden, or serve as a wildlife haven on property borders, where gardening in the prickly fallen leaves may be less of a challenge.

Erysimum capitatum, DBG.

RIGHT *Stanleya pinnata*, DBG.

BRASSICACEAE

Members of the family Brassicaceae are found throughout the world. All share the common trait of having four petals, many are fragrant, some are annuals and some long-lived perennials. Some live in infamy as aggressive weeds and pose a threat to native ecosystems from woodland to desert, but many—*Iberis*, *Matthiola*, *Alyssum*, *Aubrieta*, and dozens more—are well-known garden ornamentals. Yet within this diverse family are a good number of desirable garden plants that remain little used. Some, including North American endemics *Physaria* and *Lesquerella*, form miniature rosettes of silver leaves crowned with golden rings of bloom, well suited to rock gardens. Plants in the mustard family are abundant in the West; a few genera are shared with other steppe regions, including *Erysimum* in Asia. *Erysimum capitatum* (western wallflower) brightens sage meadows and woodland edges with branched spikes of fragrant yellow or orange; it can be difficult to distinguish from *E. asperum*. Selections of these variable species have been known in gardens for decades yet are still overlooked in the nursery trade.

In drier habitats *Stanleya pinnata* (prince's plume) raises feathery wands of golden flowers against the red sandstone cliffs of the Colorado Plateau, or tumbled gray boulders of the Great Basin and Columbia Basin. *Stanleya*, a North American endemic, often grows in soils high in selenium and is considered an indicator of such, but plants also thrive in ordinary gardens with good drainage, contributing airy spikes of yellow to the xeric garden each spring.

Opuntia polyacantha, orange form.

Opuntia aurea 'Coombes Winter Glow'.

Pediocactus simpsonii, yellow form.

CACTACEAE

Cacti conjure images of scorching deserts, but they range far beyond, from the prairies and shrub steppes well into the forested montane zones of the West. However, one stereotype of our native cacti is accurate: they do prefer sun and well-drained soils. Convergent evolution elsewhere has created many cactus-like forms, notably the euphorbias of Africa and Asia, but true cacti belong to the exclusively western hemisphere family Cactaceae. They have been distributed far and wide, sometimes with disastrous results, but most are ornamental and well behaved, and add priceless sculptural elements to the well-drained garden. Cacti spread south through the deserts and the tropics with staggering diversity, but at the far end of the continents, they shrink back again to dense well-armed cushions in the dry windy steppes of Patagonia.

In the steppes, *Opuntia* is by far the most frequently encountered genus. Dwarf clumps of *O. fragilis* may go unnoticed among low grasses, but their larger relatives are more visible, the quintessential prickly pears of old westerns. *Opuntia polyacantha* is adaptable and variable, with an extensive range from the Great Plains to California, and north into Canada. Spination varies, from the long dense "fur" of *O. erinacea* to the "spineless" *O. basilaris* and *O. aurea*. On these, only the long spines are missing. All opuntiads are well armed with tiny barbed glochids that penetrate gloves—and skin—with irritating ease. This does not deter gardeners from obsessing over them, and one of the best is *O. aurea* 'Coombes Winter Glow'. Its native range is small, spilling from the southern piñon forests of Utah across to the plateaus of northern Arizona. With a "beaver tail" form, its waxy blue-green pads sport 3-inch fluorescent magenta flowers along their edges late each spring. By winter, the pads have relaxed and wrinkled, taking on a plum-purple cast that glows in the winter garden. Drop a pad on the ground anywhere, and you have successfully propagated this easy and rewarding gem.

Small barrel-shaped cacti also grow in the steppe, including many endemics with precarious footing in their threatened habitats, but others are common, and

Pediocactus simpsonii, pink form.

The endemic
Sclerocactus glaucus
of western Colorado.

RIGHT *Echinocereus
triglochidiatus* glows in
the spring landscape.

well suited to cultivation. *Pediocactus simpsonii*, found from dry grasslands to sunny openings in ponderosa forests, is extremely variable; flowers may be shades of yellow or pink, depending on the local clone. Readily available as seedlings from nurseries, these thrive when well placed in a rock garden.

Claret cup cacti (*Echinocereus* spp.) encompass another variable group of clumping columnar cacti, including *E. triglochidiatus* and *E. coccineus*, and their many subspecies and hybrids. They are wide-ranging throughout the Southwest, common in the sagebrush and piñon-juniper zones, with some forms spilling into more desert-like habitats. When they bloom in mid-spring, they light up rocky hillsides (and gardens) in shades from peach to scarlet.

Several *Sclerocactus* species in the region are worthy of cultivation, but their genetics are still being sorted out. The endemic *S. glaucus*, *S. brevispinus*, and *S. wetlandicus* are especially vulnerable in the wild and have been placed on the federal threatened list, which has spurred continued research and protection efforts.

CUPRESSACEAE

This family includes the cypresses of art and literature, notably the tall elegant Italian cypress and its many relatives of Europe, Asia, and western North America. But the true characters of this family live in the windswept steppes of Central Asia and the Interior West of North America. Numerous species hold tightly in the ragged outcrops of drier regions, from Spain to China, and from west Texas to British Columbia. Nearly all species have value in difficult landscapes and are easily shaped into formal elements or gnarled bonsai specimens.

Utah juniper (*Juniperus osteosperma*), usually less than 15 feet tall, is notable for its typically single trunk and cloud-like form arranged on sinuous branches. Its

Juniperus osteosperma in Capitol Reef National Park.

true character is revealed in nature only after a century or two of wind, drought, and adversity. This is the formula that results in the great gnarled specimens of the western wildscape, sculpted green with fluted arms bent against red cliffs and the silent deep blue of sky. Younger garden specimens can be guided toward the same dramatic results, to be appreciated by future generations. Utah juniper is replaced by western juniper (*J. occidentalis*) in the western portions of the Great Basin and the Columbia Plateau. Western juniper is abundant—even monotonous at times—within its range, but little used elsewhere in horticulture.

ELAEAGNACEAE

The family member that leaps to mind in the West is Russian olive (*Elaeagnus angustifolia*), which is viewed with both adoration for its fragrance and durability and contempt for its invasive qualities along riparian areas. It thrives in the West, having come from nearly identical habitats in Central Asia. Several of its relatives *belong* here, however, and yet remain nearly absent in our gardens. Among the most striking of these natives is the endemic roundleaf buffaloberry (*Shepherdia*

PLANT PRIMER

193

Shepherdia rotundifolia is a touch of silver in the desert steppe landscape.

RIGHT *Ephedra viridis* adds lime-green to the landscape and shades of gold when in flower, as here.

rotundifolia). It has been challenging to propagate and resents too much care, especially excess water, but its potential is great. This densely branched shrub haunts the red canyons of the Colorado Plateau, its ghostly silver mounds huddled along cliff walls or marching down shallow arroyos. Its convex ever-silver-green leaves are tightly arranged along dense mealy stems, giving the appearance of a well-manicured silver boxwood. This native is the perfect complement to low grasses and cacti in the xeric garden; one can only hope that it becomes more readily available for use in western gardens.

EPHEDRACEAE

The joint-fir family may be monogeneric, but dozens of species of *Ephedra* inhabit dry plateaus and ravines around much of the globe. Ephedras are, oddly enough, gymnosperms, allied with conifers. Their leafless stems create a distinctive texture in the landscape, being very "desert-like" in appearance. Most are dense shrubs 2 to 6 feet tall and wide, and thickets can be much larger. Many are rhizomatous and serve to stabilize sandy soils, creating dense continuous mats. All are generally similar in detail, however, with slender stems from blue to acid-green doing all the photosynthesis, and female flowers of many species ripening into attractive red berry-like cones. Human consumption for various purposes has a long history and poses significant risks (plants contain the alkaloid ephedrine). A dozen or so species are routinely cultivated, mostly of Asian origin, the most notable being *E. equisetina*, for its blue stems and red berries. Others have untapped potential, especially the Patagonian cushion-forming species, but one of our native species deserves mention as well.

Ephedra viridis (Mormon tea) is common in the natural landscape of the West, often mixing with yucca, sage, blue grama, blackbrush, and a host of wildflowers. It is also reasonably available in nurseries. It is well behaved, seldom producing the runners that can make other species problematic. It thrives on neglect throughout the Interior West, maintaining its vivid green all year. This is a wonderful feature in the depths of winter, the bright green stems rounding out the season's palette of tan, rust, white, and blue, and adds a fine texture to the landscape all year.

ERICACEAE

The family Ericaceae is huge and widespread, its thousands of species, from large trees to tiny herbaceous plants, found on every continent but Antarctica. Despite this range, they often live in specialized habitats or with specific mycorrhizal associations. In North America the family includes the well-known genera *Rhododendron*, *Gaultheria*, and *Erica*. The common bearberry or kinnikinnick of western forests, *Arctostaphylos uva-ursi*, is not strictly our own—it is a circumboreal species of the northern hemisphere, with many regional forms throughout Europe, Asia, and North America. In the West there is a great proliferation of the genus *Arctostaphylos*, especially diverse in California and Mexico. Known as manzanitas ("little apples"), these are a hallmark of Madrean flora; all are evergreen, and several species can become small trees, famous for the satin sheen of their rust-red stems. A few species and hybrids venture into the steppe, remaining wonderfully green year-round.

Arctostaphylos patula (green-leaf manzanita) is especially durable, becoming a valuable garden specimen 4 to 6 feet tall and at least as wide. Propagation has been tricky with this species, but naturally occurring hybrids with *A. uva-ursi* have proven easier to propagate and are adaptable to well-drained garden conditions. Notable cultivars include 'Panchito' and 'Chieftain'. With a limited palette of broadleaf evergreens in the cold arid steppe, these are a welcome addition to gardens in the West.

FABACEAE

Commonly known as the pea family, Fabaceae is especially diverse in the Intermountain West and other steppe regions. Well-known genera include *Baptisia*, *Lupinus*, and *Lathyrus*, but these represent only a small portion of the family. A few weedy genera (*Medicago*, *Melilotus*, *Linaria*) from Europe and Asia are pesky invaders of American landscapes, venturing from disturbed roadsides into otherwise pristine meadows. Less well known but incredibly diverse genera include *Oxytropis*, *Hedysarum*, and especially *Astragalus*. These all have closely related species in Asia, South Africa, and Patagonia, but *Astragalus* most certainly outnumbers them all.

Little known in general horticulture, *Astragalus* is one of the most widespread and diverse genera in the family, found in all four steppe regions but

Arctostaphylos patula is a variable species.

ABOVE Inflated seed pods of *Astragalus whitneyi*.

The endemic *Astragalus coccineus* lives along the boulder-strewn base of the Sierras.

BELOW *Lupinus bakeri* has a small range but great potential in horticulture.

White stars of *Fendlera rupicola* shine against a red rock backdrop.

LEFT Flowers of *Lupinus arbustus* illuminate sage meadows in variable colors.

seldom in gardens. Species range in the thousands, but you will seldom see a landscape dominated by them. Many are local endemics, accounting for multiple entries on the rare and endangered species roster. Their soil chemistry and composition requirements can be very precise, which makes them extremely vulnerable to disturbance—and a challenge in cultivation. Nonetheless, many are stunning plants and worth the effort, and no doubt a handful will find their way into gardens eventually.

The Colorado native *Lupinus bakeri* remains virtually unknown in gardens, despite its impressive range of clear blues and sturdy habit. Garden trials continue with this species. The similar but less showy *L. argenteus* has a much wider range and only slightly less garden appeal. *Lupinus arbustus*, from the sage-covered valleys of the Northern Rockies and Columbia Plateau, is equally beautiful and nearly as unknown. By contrast *L. polyphyllus*, native to western North America, is now grown around the world, even blanketing roadsides in the Andean foothills of Argentina.

HYDRANGEACEAE

One would not generally associate hydrangeas with the arid expanses of the steppe, but the family Hydrangeaceae does have its representatives here. High elevations in the Rockies shelter *Jamesia americana* (waxflower) in damp rock crevices of the montane zone, but farther west, rugged drought-tolerant cousins venture into more open country.

PLANT PRIMER

Salvia dorrii reflects the blue skies of the West.

RIGHT Rosy spikes of *Salvia pachyphylla* play backup to the coral blooms of false yucca, DBG.

Philadelphus microphyllus (littleleaf mock orange) shares the characteristic of delicious fragrance with its more lush eastern and Asian counterparts, but the plant is more compact, with tiny leaves; it grows on canyon sides and in rocky crevices of the Colorado Plateau and the Great Basin. To the northwest, it is replaced by *P. lewisii* (Lewis' mock orange), which finds its footing even in the splintered surface of ancient lava flows; this likewise fragrant and durable shrub sports the classic pure white blooms and good habit without the water needs of more common species. In fact, 'Cheyenne' (a Plant Select promotion) hails from that Wyoming city's decommissioned USDA Research Station, where it survived decades of neglect after the taps were turned off.

The closely related *Fendlera rupicola* occupies sandy flats and ledges in much of the Southwest, especially the Colorado Plateau. Four white spoon-shaped petals have a star-like quality, and rare specimens have hints of pink. These tough deep-rooted shrubs respond well to pruning, or will form a multi-stemmed dome about 5 feet tall with gnarled trunks over time. Ideally suited for use as a specimen, or a naturalistic backdrop to the xeric garden.

LAMIACEAE

In our region the most showy members of this family are the salvias. While other regions boast dozens of herbaceous species, our stars are small shrubs generally under 2 feet tall. *Salvia dorrii* ranges throughout the Great Basin and Columbia Basin, brightening the thin soils of rocky outcrops where it sinks woody roots into deep crevices. Soft to bright blue flowers emerge from shorter wine-tinted calyces that persist after flowering. Small silvery leaves reflect the intense sun and

Allium platycaule from the Great Basin performs well in gardens.

persist through the winter. Variation can be considerable; there is potential for selection of superior forms, though even average forms are striking in xeric rock gardens.

Salvia pachyphylla is really a borderline steppe plant, residing along the southern and western edge of the Great Basin, in territory that may more correctly be classified as Mojave Desert, but it is close enough—and important enough—to warrant a mention. Similar in most respects to *S. dorrii*, it raises the ornamental bar, being larger in every respect. The flowers are showy and long-lasting, and even the persistent rosy calyces extend the display as they turn tan in the fall. Again, a high degree of variability calls out for selection of the finest forms. Both species require well-drained soil and full sun, thriving in unwatered gardens in much of the West.

LILIACEAE

Many plants have spent time under the umbrella of the family Liliaceae. Further study has sent some members off on their own, including to Agavaceae, Asparagaceae, and Melanthiaceae. Wild onions (*Allium* spp.), for another example, are sometimes placed in Alliaceae or Amaryllidaceae; in any case, they are important bulbs in the West, with many dozens of species. *Allium acuminatum* (tapertip onion) grows throughout the region; it is often seen flowering by the thousands in loamy clay soils on the steppe and can paint acres of land vibrant pink in early summer. Many similar species can be found, but sheer numbers make this among the most conspicuous. *Allium platycaule* (broadstemmed onion) is found in the northern Great Basin; it can form vigorous clumps, adding vibrant color as a choice and easy rock garden species.

Noteworthy in the arid West, and still in Liliaceae by most accounts, are the

Stark desert steppe seems an unlikely habitat for such a delicate flower as *Calochortus flexuosus*.

graceful sego or mariposa lilies, *Calochortus*. Related to and resembling the true tulips of Central Asia, these are the wild "tulips" of the West. There are many species and subspecies, especially diverse in the many habitats of California. But in the steppes, a handful of species tough it out in well-drained soils, often appearing en masse across boulder-strewn fields or canyon slopes in sandy soils. The slender bulbs of the sego lily are less than an inch across, and in their native soils may be 5 inches deep or more. Flowers of most species are conspicuous, 1 to 3 inches across, with three simple petals intricately patterned near their centers, as if hand-painted with tiny brushstokes. Most also have fine hairs near the center (and even extending over much of the petals, in some species). Calochortus are a challenge to establish and slow to multiply, but neglect is often the answer—they truly prefer dry conditions and deliberate disregard. Tissue culture micro-propagation has made many species readily available and inexpensive, but untold potential is yet to be realized with these gems of the steppe.

Calochortus flexuosus is found in the southern Great Basin and Colorado Plateau. Flowers range from nearly pure white to a rich rose-pink, with most somewhere in between. The centers are golden to orange, usually lacking the distinctive patterns found in other species. Its leaves have a conspicuous curl, and even the flower stems lean and twist, each bearing up to three flowers. These are seldom cultivated but have great potential for arid gardens, and the variations in flower color and habit offer tempting possibilities for selection. One of the hardiest species is *C. nuttallii*, common in the eastern Great Basin, the Colorado Plateau, and on into much of the Rocky Mountain region, but uncommon in cultivation. White is the dominant color, but occasional pink populations are especially striking, as all have dark spots or crescents painted on the interior of the petals against the golden centers.

MALVACEAE

This family has especially diverse members, from tiny herbs to giant tropical trees. Well-known members include the common hollyhock, tropical hibiscus, and cotton. All share the trait of having five petals. In the West, *Sphaeralcea* is especially common, with dozens of species scattered throughout the steppe, deserts, and mountains; nearly all its members are worthy of cultivation, offering dominant colors of orange to salmon, with the occasional pink or white. Three noteworthy species are *S. grossulariifolia*, *S. munroana*, and *S. parvifolia*. All light up the landscape in rich salmon-orange. *Sphaeralcea munroana* and *S. grossulariifolia* occur from the Columbia Basin throughout the Great Basin, but *S. grossulariifolia* ranges further into the Colorado Plateau, distinctive with its deeply divided "goose-foot" leaves. *Sphaeralcea parvifolia* is most abundant in the Colorado Plateau and southern Great Basin. All three species may have the appearance

Lewisia rediviva finds the harsh conditions of Brown's Park, Colorado, especially hospitable.

LEFT *Sphaeralcea parvifolia* against the red sands near Utah's Fisher Towers.

and habit of small shrubs, maintaining a branched woody base with flowering stems that die back in the winter. *Sphaeralcea parvifolia* is less likely to have leaves among its flowers, presenting a dense crown of orange that will stop you along the road or trail and demand that a photo be taken!

MONTIACEAE

The Montiaceae were formerly included in the family Portulacaceae, and they do share common traits of succulent leaves and stems, and a degree of drought tolerance. This former classification linked them more closely with familiar (even weedy) plants like purslane and the ornamental portulacas and talinums. Separated now, Montiaceae still includes *Calandrinia*, *Claytonia*, and *Phemeranthus*, to name a few, all with strong links to South America. The stars in the intermountain steppe are in *Lewisia*, with more than a dozen species, some rare endemics with very limited distribution. Most widespread is *L. rediviva*, which is treasured by rock gardeners everywhere. Plants can thrive in well-drained conditions. Out of bloom, their small succulent green rosettes are easily overlooked; in spring, stunning white to pink blooms seem to appear out of nowhere on dry rocky plains and gravelly slopes.

NYCTAGINACEAE

Several obscure genera in this family occupy the Americas, one being *Abronia*. Most abronias have some fragrance, especially *A. fragrans* of the Great Plains and

Mirabilis multiflora on Colorado's Uncompahgre Plateau.

the Colorado Plateau; even more showy is *A. nana*, with its pink umbels scattered across sandy flats and washes of the Southwest. Well-known garden plants in this family include *Bougainvillea* and *Mirabilis* (four o'clocks), both of South American origin. The star of the North American steppe is *M. multiflora* (desert four o'clock), scattered from the southern Great Plains to the southern Great Basin and southward into Mexico. This tough perennial sports bright purple flowers each afternoon for most of the summer. It thrives—even requires—neglect once established. The thick tuberous root makes it bulletproof against drought, and any excess of water transforms it from a tidy 3-foot mound into a sprawling 8-foot tangle of leaves. This is a perfect companion plant in a xeric garden of cacti, agave, yucca, sagebrush, mountain mahogany, piñon pine, and dozens more.

ONAGRACEAE

Members of this family have undergone numerous name changes over the years. Fortunately, that does nothing to diminish their beauty or usefulness. Common members of this family include fireweed, evening primrose, California fuchsia, and the related true *Fuchsia* of South America. Several genera are shared between continents, notably *Oenothera* and *Epilobium*. All have flowers with four petals, and many have fine hairs on leaves or stems.

Calylophus lavandulifolius is among the best of the several species of sundrops that venture from the western Great Plains to the Colorado Plateau and

LEFT *Calylophus lavandulifolius* on rocky slabs near the Gunnison River, Colorado.

Oenothera caespitosa, DBG.

Epilobium canum growing to perfection in Kendrick Lake Park, Lakewood, Colorado.

central Nevada. Well-drained, dry rocky soils will suit it; in the wild it is found on thin soils that accumulate on fractured shelf rock, or in open expanses between tufts of grass and sagebrush. It is excellent in a xeric garden; its pure golden flowers open during the day (unlike those of many of its relatives) and continue over a long season.

Many epilobiums inhabit the West, often in meadows among ponderosa pines, on the edges of sage-filled meadows, or on canyon slopes among grasses and shrubs. Most are from farther west, but *Epilobium canum* rings the fringe of the Great Basin from California to Utah, Wyoming, and southern Idaho. With small leaves and a slowly running habit, this species is well suited to a xeric border or large rock garden, irresistible to hummingbirds and a great companion to late-blooming xeric asters. Its brilliant red-orange trumpet flowers last from late summer till frost, though some cultivars have an earlier season of bloom.

Oenothera caespitosa (tufted evening primrose) is widespread, from the plains of Canada to Texas, and west to the Columbia Basin. It thrives in sandy soils of the sagebrush steppe, with fuzzy gray-green rosettes of leaves, opening its gleaming white flowers in early evening; flowers fade to pale pink as they close in the heat of the next day. Many similar species are found in dunes and sandy washes of the West.

OROBANCHACEAE

Members of this important family used to be included in Scrophulariaceae, alongside penstemons, verbascums, and diascias, but a host of genera have recently taken up taxonomic residence here. Still, it houses a respectable, if underused, lot of ornamentals with strong ties to the floras of South Africa, Asia, and South America. Most in this family are fully or partially parasitic, with flowering spikes emerging from bare ground, garnering sustenance from the roots of nearby plants. Flowers can be striking and colorful. They usually have specific hosts, and some are found only in association with sagebrush. This makes cultivation of many species very challenging; few are found in cultivation.

Notable in the West is *Castilleja* (Indian paintbrush). Species range from hot deserts to damp subalpine meadows, in shades of pink, yellow, and red. They are known to be hemiparasitic, but in the company of the right plants, they can perform well in xeric gardens. Steppe species appreciate perfect drainage. *Castilleja integra* is one of the most common in the Southern Rockies and Great Plains, and into the Colorado Plateau. It is often sold in containers along with blue grama and fringed sage, to encourage the hemiparasitic relationship with these species. Once established, it can be long-lived, and its long flowering season makes it a treasure to gardeners and hummingbirds alike. The closely related *C. linariifolia* is a larger plant with a distribution extending throughout the Great Basin and

Castilleja angustifolia var. *dubia,* unusual peach-colored form.

Castilleja integra.

Castilleja linariifolia.

Columbia Basin; it may reach 2 feet or more in height, often leaning on neighboring plants for support. A contrast in stature is the short tufted *C. angustifolia* var. *dubia*, usually with flame-red flowers, but there is variation. It ranges widely throughout the West but is nowhere more striking than against the rusty soils and cliffs of the Colorado Plateau. Once established, it is a highlight in a well-drained rock or desert garden.

PINACEAE

Pines dominate many of the semi-arid portions of the northern hemisphere. In western North America, although the dominant forest mix is piñon pines and junipers, ponderosa and lodgepole pines have the widest range, forming extensive forests just a bit higher in elevation than most sagebrush steppe, their habitats intergrading, depending on exposure and precipitation patterns. In the Colorado Plateau the common species is *Pinus edulis*, with similar species replacing it farther south into Mexico. In the Great Basin the dominant species is *P. monophylla*, distinct with its single needle per fascicle. Selected blue forms grow into stout pyramids in cultivation. Both species produce the famed piñon nuts that are so important for wildlife and still provide sustenance to Native Americans of the region. Over time both species are capable of becoming gnarled and picturesque specimens. The scent of burning piñon wood is a trademark experience of traveling through towns and villages of the Southwest.

Pinus edulis assumes many forms.

PLANTAGINACEAE

Revisions to the classification of plants can lead to head-scratching and frustration, but molecular and DNA studies do not lie. Thus we now include one of the

Penstemon arenicola in the sandy flats of Wyoming.

Penstemon utahensis near Moab, Utah.

Penstemon palmeri sweetens the air of Great Basin National Park.

most important genera of the West in Plantaginaceae rather than Scrophulariaceae: *Penstemon*. This process is still in flux, so expect further surprises. Rest assured, however, that *Penstemon* remains in good company, an unlikely cadre that includes *Veronica*, the dazzling *Ourisia* of Argentina, and a cast of thousands.

Penstemons are a frequent and welcome component of steppe habitats, from alpine tundra to the sizzling Sonoran Desert. The low-growing *Penstemon arenicola* lives primarily in sandy soils of Wyoming and the Uinta Basin but performs well in rock gardens with good drainage. *Penstemon mensarum* is a western Colorado endemic, its 18-inch stems stacked with electric cobalt and purple flowers that ensured it a place in the Plant Select introduction program. In the Colorado Plateau the coral-red spires of *P. utahensis* glow against red canyon walls as they do in xeric gardens. From New Mexico to Nevada and Idaho *P. eatonii*, the firecracker penstemon, ignites cliffs and roadsides in mid-spring with scarlet spikes, providing a first meal to migrating hummingbirds headed north. A similar, more extravagant show is provided by *P. barbatus* later in the season; its 2- to 5-foot wands of red trumpets last for many weeks, extending from southern Utah and Colorado well into Mexico. The large flowers of *P. palmeri*, found from New Mexico to the Columbia Basin, resemble snapdragons in shades of soft to deep pink on stems up to 5 feet tall; unique in the genus, this species exudes the sweet fragrance of grape soda.

This is only a small sampling. While some of the scores of species and

Sporobolus airoides, DBG.

LEFT Shimmering *Hordeum jubatum*, DBG.

subspecies are rare endemics, most have great potential—diverse enough to find a place in any well-drained garden and well worth seeking out.

POACEAE

Grasses are common the world over in nearly every habitat, but they reach their climax in the great steppes of the world. They are less dominant in the sagebrush steppe, usually acting as part of the mix rather than the main event, but there are hundreds of species, and in some habitats they do dominate, especially where moisture is present, or in the wake of recent fires. Introduced grasses (e.g., cheatgrass) from other steppe climates have also found an easy foothold here and now dominate thousands of square miles. But the native grasses persist, and often bind entire landscapes together, sweeping seamlessly from one habitat to the next.

Hordeum is common in the intermountain steppe, with species in Patagonia, Africa, and Asia as well. The important crop we know as barley belongs in this genus. Each year I note the beginning of summer when I see the nodding flowers of foxtail barley (*H. jubatum*) shimmering in the wind along roadsides and in low swales. This grass is seldom used, but it adds a graceful movement to the garden that is unrivaled. It will reseed abundantly, but if trimmed short before seed disperses, it will rebloom again later in summer for a repeat show. In some years a third show is possible, and reseeding is kept to a minimum.

In lower swales or playas, *Sporobolus airoides* (alkali sacaton) forms bright green tufts against chalky alkali flats, yet it is surprisingly well adapted to cultivation. Smooth arching leaves form lush clumps to 24 inches high, topped with an airy froth of blooms over a long season. As with most ornamental grasses, cutting back each spring spurs quick regrowth.

PLANT PRIMER

Hesperostipa comata,
needle-and-thread.

RIGHT Blue grama with
seeds in winter.

Bouteloua gracilis (blue grama) is common on the Great Plains but also appears in various forms and subspecies across most of the West. It seldom grows more than 12 to 16 inches tall. In a garden it is a well-behaved clumping grass, resistant to drought and providing the perfect foil for hundreds of wildflowers and drought-adapted plants. Curled seedheads add interest all winter.

Achnatherum hymenoides (Indian ricegrass) also has a wide range, from the Mississippi to the California coast, but throughout the Intermountain West it is commonly associated with sandy windswept bajadas or sandy canyon walls, where it is often the only grass present. It survives significant drought, and its zigzag inflorescence lends an airy quality to a desert garden.

Stipa is equally widespread with hundreds of species, though some now fall into the genus *Hesperostipa*. *Hesperostipa comata* (needle-and-thread) is a common component in sandy soils from the Great Lakes to the Columbia Basin and south into Mexico, but it can dominate large expanses in the steppe, accompanied by evening primrose, sunflowers, cleomes, and other sand-loving plants. Its long silky awns and sharp-pointed seeds (hence the common name) wave in the slightest breeze, setting the world in motion. Equally ornamental and found in similar habitats of the Southwest is the closely related *H. neomexicana*, with longer feathered awns up to 4 inches. These are welcome additions to a xeric garden with such companions as yuccas, agaves, penstemons, and cacti. Several Asian and South American *Stipa* species with similar qualities and habitats are also cultivated.

POLEMONIACEAE

The phlox family is centered in western North America, with some members extending south into the Andes and a handful in Europe and Asia as well. Common tall garden phlox is familiar to most gardeners, but in the West, low-growing *Phlox* species dominate, forming dense cushions or mats on well-drained slopes or rocky soils; some species in exposed sites will capture windblown dust in their crowns, slowly growing up into living bowling balls scattered across barren flats. Most of the dozens of species, including many endemics, are covered in spring with typically white (but sometimes pink or blue) flowers. *Phlox hoodii* has a wide range, extending from the western Great Plains all the way to the mountains of California and north to Canada and Alaska, especially in the more arid interior; it performs well in rock gardens with good drainage. Less common though with a similar range is *P. diffusa*, usually with pink flowers. *Phlox grayi* boasts a range of rich pink flowers. Selections especially suited for the rock garden have been made from many species, but there is potential for selection with wider use in mind.

Also in this family is *Ipomopsis*. In the steppe, *I. aggregata* is an especially conspicuous biennial in the montane zones and throughout the piñon-juniper

Eriogonum ovalifolium,
soft pink form.

Eriogonum corymbosum
forms creamy cushions.

Eriogonum umbellatum var. *aureum* Kannah Creek
with *Penstemon pseudospectabilis*, DBG.

and sagebrush habitats. Ferny-leaved rosettes shoot stems up to 2 feet in their second year, with flared tubular flowers in shades of coral-red to pink, irresistible to hummingbirds and wonderful in a xeric garden.

POLYGONACEAE

From herbaceous perennials to vines and trees, members of the buckwheat family are found worldwide, from the tundra to the tropics. Familiar plants in cultivation include agricultural buckwheat, rhubarb, polygonums, persicarias, and Mexican coral vine. In the West, the larger genera are *Rumex*, many of which are considered weedy plants (some introduced from Eurasia), and the wild buckwheats of *Eriogonum*.

Eriogonum species occur in the hundreds throughout the steppe. They are a complicated group, with a slew of rare endemics in isolated habitats, but a few have garnered horticultural interest—even fame—and many more are candidates in waiting. *Eriogonum umbellatum* (sulphur buckwheat) has dozens of varieties, and great variation within those varieties. Low mats of evergreen leaves take on reddish shades in winter. In spring and summer, branched umbels support dense spherical clusters of tiny flowers that are impressive en masse, ranging from greenish white to rich yellow, and aging to shades of pink or rusty red. Nearly all are

Aquilegia micrantha clings to vertical sandstone seeps in remote canyons.

ornamental, and several superior garden selections have been made; one of the best of these, from var. *aureum*, is Kannah Creek, promoted through Plant Select, but there is room for more. *Eriogonum ovalifolium* is a variable species, forming dense cushions of silvery leaves, topped in spring with single-stemmed spheres of flowers in cream, yellow, or soft or deep pink. These are treasures for well-drained rock gardens or xeric borders, but little selection has been done. When flowering, *E. corymbosum* can be a giant by comparison, forming large wiry domes 2 to 3 feet across. Creamy flowers by the thousands age to rust or pink, and again, there is room for horticultural selection. All require good drainage and a degree of neglect to maintain their compact forms.

RANUNCULACEAE

This diverse family includes far more than buttercups. Other familiar garden plants include anemones, delphiniums, monkshood, hellebores, and clematis. Also residing here are the columbines, universal favorites evoking visions of high mountain meadows; but *Aquilegia* species occur on the steppe as well and even in the desert, where moisture and drainage suit them. *Aquilegia desertorum* is a transitional species, found in the edges of ponderosa pine forests among oaks and sage in canyons of the southern Colorado Plateau; its robust red and yellow flowers on sturdy stems are a wonderful addition to gardens with moderate water and good drainage. *Aquilegia micrantha* occupies a similar range, though its habitat is

LEFT *Clematis hirsutissima.*

RIGHT *Pulsatilla patens* is among the earliest spring flowerers.

more specialized. You will not find this on the forest floor or open meadows; rather, it clings to vertical damp seeps in sandstone walls with maidenhair fern and mosses. Delicate flowers in soft shades of lemon, pink, and white nod on wiry stems against the red cliffs. Not truly a plant of the steppes, *A. chrysantha* occupies similar canyon habitats along the southern fringe of the region, and into the Chihuahuan and Sonoran deserts; it is especially robust and performs well in cultivation. Most columbines hybridize freely, resulting in offspring with variable characteristics.

Clematis are found in all four steppe regions as herbaceous perennials or vines. *Clematis ligusticifolia* is abundant throughout the West, scrambling up trees and along dry riverbanks, with scores of tiny white flowers followed by fluffy seedheads in fall. Many species look similar, but most are a bit too rambunctious for use in the average garden; however, several less rampant non-vining species inhabit the West, from the Great Plains to the Columbia Plateau and south to Arizona and Mexico. Among them is *C. hirsutissima*, which is a delight in the garden and very different from its large-flowered cousins. It has a wide range, appearing in grasslands, sage meadows, and open pine forests. Toughing it out in dry conditions, its nodding blue bells on foot-long stems keep most people guessing. It thrives in xeric gardens with infrequent watering and is long-lived.

Pulsatilla patens (pasqueflower) is common from the northern Great Plains to the western ranges of the Rockies and south into New Mexico, thriving in openings in the ponderosa forests and at the edges of sage-filled meadows. Its close Eurasian relative, *P. vulgaris*, is more commonly spotted in gardens but is no more beautiful than our native species, which is one of the first to bloom each spring.

A hardy blue-flowering ceanothus would be a welcome addition to horticulture.

LEFT The endemic *Ceanothus greggii* var. *franklinii* displays purple and blue flowers on a hardy desert shrub.

RHAMNACEAE

The famed genus of this family in the West is *Ceanothus*. In California, where they come in a dazzling array of blues and purples, they are known as summer lilacs (though they are not related). The few hardy species that venture into the Intermountain West—*C. fendleri*, *C. velutinus*, *C. greggii*—have smaller flowers, usually creamy white, on small shrubs. *Ceanothus velutinus* (snowbrush ceanothus) is found along the fringes of the northern steppe, especially abundant in parts of the Columbia Plateau, on grassland edges, among sage, and well into the lower ponderosa pine forests where sunny openings allow. Creamy white flowers resemble small lilacs from afar. It is a broadleaf evergreen with large leaves—a rare feature of steppe plants generally—and though challenging to establish in cultivation, it is worth seeking out for a well-drained naturalistic garden. Scattered in remote locations of the Colorado Plateau is the rare blue *C. greggii* var. *franklinii* (desert ceanothus). This slow-growing intricately branched shrub persists only on windy caprock ledges, where windblown sands accumulate, and is inconspicuous among similarly stunted mountain mahogany and junipers—until spring, when it is transformed under puffs of fragile flowers, from soft pink to royal purple, an unlikely vision against the dusty soils of the canyon rims. Not yet in cultivation, this stunning variety may have potential in horticulture or in future breeding programs.

PLANT PRIMER

Purshia stansburiana has larger whiter flowers.

LEFT *Purshia tridentata* in a Utah canyon.

ROSACEAE

The rose family needs no introduction, but many people are surprised by its size and diversity. Besides ornamentals, the family includes a wealth of edibles that have shaped the human experience: apples, pears, almonds, peaches, cherries, and strawberries, to name a few. Many of its perennials and shrubs have a firm place in regional gardens, and many lesser-known species from the steppe are especially valuable here in the Intermountain West.

Amelanchier utahensis (Utah serviceberry) blooms early, billowing white against the lime-green of new aspen leaves. These slowly suckering shrubs can reach 10 feet in height, and some display good fall color. The dark berries are important for wildlife and foraged by resourceful people as well. Superior forms could be selected for horticulture.

Fallugia paradoxa (Apache plume) also has white blooms, like small single roses, but occupies lower elevations of the steppe, extending well into the grasslands and desert country to the south and west. It flowers over a long season, and seeds mature into fluffy plumes that persist into the fall. Some forms sucker with abandon, some boast larger flowers or pink plumes; such variety presents another great opportunity for selection.

Cliffroses (*Purshia* spp.) too are variable enough that better forms could be selected for garden use. The most common species are *P. stansburiana* and *P. tridentata*. The latter has the widest range, extending throughout the Columbia Basin and into British Columbia. The flowers are not large, but they cover the irregularly branched shrub in simple creamy to yellow blooms in mid-spring, lending a warm glow to the rugged landscapes it inhabits. *Purshia stansburiana* has larger creamy flowers in similar profusion masking the tiny lobed leaves. It brightens the canyons and dry slopes of the Colorado Plateau and eastern Great Basin, its cascades of bloom alive with foraging honeybees. Peeling bark and a gnarled habit create a highlight in any xeric or native landscape. To the southeast, in the Chihuahuan grasslands of Mexico, the closely related *P. plicata* sports rich pink blooms and presents yet another untapped opportunity for selective breeding.

SAPINDACEAE

Maples are familiar trees in much of the northern hemisphere, ranging throughout forested areas of North America, Europe, and Asia, which holds the greatest diversity of species. Several small maples inhabit the Intermountain West. *Acer glabrum* (Rocky Mountain maple) is an understory shrub or small tree in many western forests, mingling with pine and fir. Especially useful in western gardens is *A. grandidentatum* (bigtooth maple); with its epicenter in the Wasatch Range,

Boxelder graces a narrow canyon in southern Utah.

Heuchera rubescens,
pink form.

it is found from sheltered ravines in west Texas to the grassy foothills of Utah and Idaho. *Acer negundo* ssp. *interius* (boxelder) casts welcome shade in canyons and riparian areas.

SAXIFRAGACEAE

The diminutive plants of the saxifrage family are cherished by rock and alpine gardeners. Most members of this family like plentiful moisture, but a few genera are adapted to more arid conditions. In particular, a handful of endemic heucheras have found their niche in rock crevices and shaded slopes of the West, their clustering rosettes of scalloped evergreen foliage rooting deeply, often in full view of prickly pears, yuccas, and sagebrush. They are a variable lot, and selected forms have been used extensively in breeding programs for select flowering or foliage characteristics. *Heuchera rubescens* is widespread, from west Texas to the southern Columbia Basin; the individual blooms, albeit tiny bells, typically white but sometimes soft pink, can be showy en masse. *Heuchera parvifolia* is a Rocky Mountain and Great Basin species, found from the foothill forests, where sagebrush intermixes with ponderosa pines, and up to the subalpine zone; its tiny flowers are cream to pale yellow on tall stems. *Heuchera cylindrica*, similar but with more substantial flowers, larger and brighter white, illuminates shaded cliff walls of the Columbia Basin and northern Great Basin.

Heuchera cylindrica,
Steptoe Butte, Washington.

THE
PATAGONIAN
STEPPE

MIKE KINTGEN

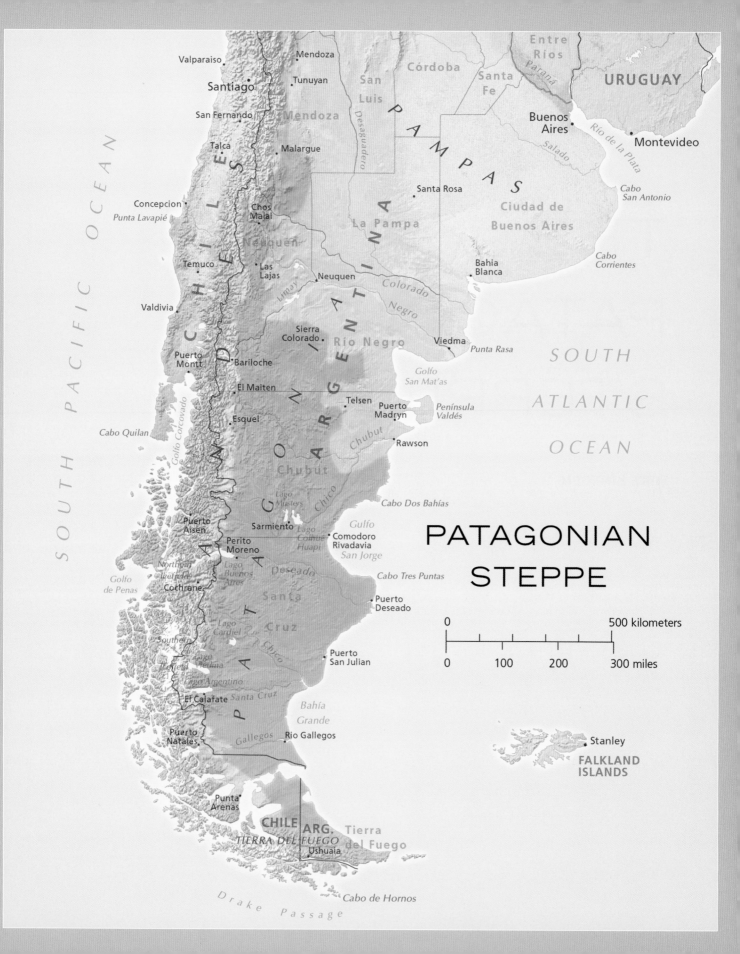

PATAGONIAN
STEPPE

MENTION THE WORD "PATAGONIA" and many things spring to mind: jagged granite spires jutting into a vast sky, incessant wind, barren plains, maybe even an outdoor clothing company. All are elements of a region in southernmost South America that touches both Argentina and Chile. The region of Patagonia, from 39°S to the very tip of Tierra del Fuego at 55°S, encompasses remarkable diversity, ranging from Valdivian rainforest on the coast of Chile, up and over the southern Andes, down through the semi-arid steppe of Argentina, and on to the penguin-studded shores of the South Atlantic. For many, Patagonia represents (as the "wild west" did in North America) the final South American frontier, where pioneering spirits go to be free. It holds its own romantic draw and continues to attract visitors to explore and marvel at its scenic beauty and fragile ecosystems.

The region of Patagonia and the Patagonian steppe, however, have different outlines. The Patagonian steppe is a smaller area within Patagonia proper, mainly in Argentina, covering about 730,000 square kilometers. Small pieces of the true steppe extend into southern Chile, where the rain shadow of the Andes creates suitable conditions. It also ranges north along the base of the Argentine Andes to the province of Mendoza, where it forms the transition zone between the Monte to the east and the Andean zone to the west. Although it shares defining characteristics with the other steppe regions (large mountain range blocking moisture, semi-arid climate, warm summers and cold winters, high average wind velocities), Patagonia is the least horticulturally successful of the four; very little of its true steppe flora is represented in cultivation.

OVERLEAF Cerro Fitz Roy, the inspiration for the Patagonia clothing company's logo, rises above the steppe near El Chaltén.

The steppe meets the base of the Andes near Esquel, Chubut.

Geography and Climate

The Andes run the length of the South American continent from Venezuela to Tierra del Fuego, forming a barrier to moisture from the west and also moisture from the east in the latitudes closer to the equator (hence the deserts of coastal Peru and Chile and the wet Amazonian basin to the east of the Andes). In Patagonia it is this barrier to the predominant westerlies off the Pacific that keeps moisture from reaching the steppe, creating a lush temperate rainforest on the moist west side of the Andes and a semi-arid region to the east. A similar phenomenon can be observed on the west coast of North America, where the Pacific coastal forests of Oregon, Washington, and British Columbia give way to steppe east of the Cascades and coastal mountain ranges. Occasionally from Lanín National Park and south, low openings within the Andes permit moist mild Pacific air to permeate into Argentina, allowing the Valdivian rainforest to cross the Chilean-Argentine border; Nahuel Huapi, Lago Puelo, and Los Alerces national parks all have examples of temperate rainforests. Once across the international border the moisture-laden winds quickly shed their moisture, and the precipitation gradient can be extreme, with rainforest at the west end of the valleys, in the aforementioned national parks, and semi-arid forest steppe ecotone at the eastern entrance to the valleys. Nahuel Huapi National Park is an excellent example of the full gradient, with Puerto Blest a Valdivian forest and Dina Huapi, at the east end of the lake, experiencing semi-arid conditions. This entire transition occurs in about 50 miles.

Another similarity of the Andes to the Cascades of the American West is the presence of stratovolcanos: both regions are on the eastern side of the Ring of Fire, a circular line of volcanoes lining the shores of the Pacific. So the Andes, a complicated mountain range of both uplifting and volcanism, not only block moisture but have also affected drainage patterns and provided the material for many of the soils on the steppe. The Andes begin to lose elevation south of 34°S and by the time they reach Patagonia, the highest peaks are between 11,000 and 13,000 feet (mainly stratovolcanos or areas of abrupt uplifting, such as Cerro Fitz Roy or the San Lorenzo Massif), with many in-between peaks topping out at 6000 to 8000 feet, a pattern very similar to the Cascades.

In general the steppe tends to slope gently from the base of the Andes east to the Atlantic coast, varying in elevation from 1000 meters above sea level to less than 400 meters in valleys and closer to the ocean. The region, however, is far from flat, with a complicated series of mesas, small sierras, and expansive basaltic plateaus, mesetas, and canyons breaking the monotony. All these changes in topography create a variety of microclimates—and vegetation. In Río Negro, Chubut, and Santa Cruz, large areas of basaltic flows, similar in many respects to those of Washington, Idaho, Oregon, and Nevada, have formed small to vast tablelands. The "small sierras" qualify as small only in relation to the size of the Andes themselves: in several places, these independent ranges reach 1500 to 1800 meters above sea level, high enough to influence precipitation and temperature, and in parts of the sierras in Chubut and Río Negro, the flora is alpine (or pseudoalpine) in appearance.

MAJOR RIVERS

Large rivers originating in the Andes tend to flow through canyons and valleys toward the Atlantic, but in some areas the Continental Divide between the Pacific and Atlantic oceans sits subtly on the steppe, and the drainage from the east slope of the Andes and far western part of the steppe actually flows into the Pacific Ocean through the crest of the Andes. This is most noticeable in the Lago Posadas area, where a crest around the basin keeps water from draining to the east. From north to south, eight major rivers—Colorado, Negro, Chubut, Deseado, Chico, Santa Cruz, Coyle, and Gallegos—drain toward the Atlantic. As in all semi-arid regions of the world, these are ribbons of life to the steppe.

The Río Colorado is born in the Andes of Mendoza and stays a course southeastward to the Atlantic via the Bahía Blanca; it is often regarded as the official dividing line between Patagonia and the rest of Argentina.

The Río Negro, formed from the union of the Río Lamy and Río Neuquén,

The highly regarded endemic genus *Benthamiella* reaches its center of diversity within the basaltic tablelands of Chubut and Santa Cruz.

The Río Lamy, northeast of Bariloche.

flows east and northeastward toward the Bahía Blanca. Home to the largest manmade reservoirs in Patagonia, it is a source of hydroelectricity and irrigation waters for one of Argentina's major fruit-growing regions. Humans have found that many of the world's steppe climates are excellent for growing peaches, plums, apricots, and apples, all of which originated in or near the boundary of the great Eurasian steppe. Many of the world's best wine-producing regions are not far from the boundaries of steppe and Mediterranean climates. Argentina's Mendoza wine region brushes the transition from steppe to Monte and desert. South Africa's fruit and wine country overlaps the line between the Western Cape and the interior steppe regions of the Karoo. The wineries of British Columbia, Washington, and Oregon actually meld into the sagebrush steppe of the Columbia Plateau. French and Spanish wine regions blur the climatic line between steppe,

The famous vineyards of Mendoza stretch below the Andes.

Mediterranean, and humid temperate.

Arising in the Andes of Chubut province, the Río Chubut and its conspirator the Río Chico join forces on the steppe to create one of the longest and most scenic river valleys in Patagonia. For much of its length the Río Chubut winds its way through the volcanic and sedimentary sierras of central Chubut. Beyond these rocky constraints, it breaks free to form a welcoming valley as it heads toward the Atlantic, south of Puerto Madryn.

Hidden underground for much of its length, the Río Deseado ("desired river") charts a course across the steppes of Santa Cruz province. It most likely drained the larger Lago Buenos Aires of the Pleistocene, when glaciers prevented its current exit to the Pacific via the Río Baker.

Tapering in width along with South America, the Río Chico in Santa Cruz province has a comparably short journey from its glacial birth in the southern Andes to the Atlantic. Too far south for cultivation of crops, the valley it creates is home to sheep farms; and, more recently, the rich wildlife of the river's wetlands attracts ecotourists from around the globe.

The Río Santa Cruz, the namesake river of Argentina's southernmost mainland province, arises in the largest ice cap in the southern hemisphere outside of Antarctica; from glacial Lago Viedma and blue Lago Argentino, it gathers strength for its short trip to the Atlantic. The Río Coyle drains the Meseta de las Vizcachas and mountains south of Lago Argentino. And finally, Patagonia's eighth and southernmost major river, the Río Gallegos, drains the low mountains south of the Meseta de las Vizcachas, near the Chilean border.

Nothofagus forest marks the transition from steppe to the subantarctic province. Beeches need higher precipitation and atmospheric humidity, which restricts their growth to such areas.

LAKES

Water is a tremendous presence in Patagonia. From southern Neuquén to Tierra del Fuego, large lakes dominate the major valleys of the Andes, draining either into the Atlantic or the Pacific, depending on the location of the Continental Divide; the largest of these are Lago Nahuel Huapi, Lago Buenos Aires, and Lago Pueyrredón. Although they are generally much smaller, similar glacial lakes are found along the eastern slope of the Cascades in Washington State and along the front of the Rockies abutting the steppe in Montana and southern Alberta.

Lakes can also be found out on the steppe, far from the base of the Andes. Lago Musters and Lago Colhué Huapi are excellent examples, and even Lago Cardiel sits some distance from the main Andean chain. These lakes are not necessarily glacial in origin; like their counterparts in western North America (the Great Salt Lake, Utah Lake, the Klamath lakes), they may be remnants of the much larger lakes of Pleistocene time.

ECOSYSTEMS

The Patagonian steppe is differentiated from surrounding regions by its semi-arid moisture regime, warm summers, and cold winters. An increase or decrease in either temperature or rainfall (or both) would result in a different ecosystem. If one increases temperature, the result is the Monte, an ecosystem dominated by

Austrocedrus chilensis, the most adaptable of the southern Andean trees, tolerates more drought and lower atmospheric humidity than any other tree species.

spiny shrubs similar to those found in the deserts of southwestern North America; the Monte transitions into steppe along the steppe's entire northern boundary, on the imaginary line where the average temperature climbs above 13°C. Increase precipitation and decrease temperature, and the subantarctic (Andean Patagonian forest) province takes hold; dominated by *Nothofagus* (southern beeches) and *Austrocedrus chilensis*, it marks the transition from steppe to the Valdivian forest to the west. A marked decrease in temperature and increase in elevation creates the High Andean province, which is similar to high-elevation alpine regions around the world. Decrease precipitation below 7 to 10 inches, and true desert results. The gradation from steppe to desert in the provinces of Chubut and Santa Cruz is especially hard to define; each author has a different opinion.

TEMPERATURES

Patagonia's relatively small and narrow landmass, in comparison to corresponding steppes in North America and Asia, means that the moderating influences of three seas have a greater effect upon it, and Patagonia therefore never experiences the temperature extremes found at similar latitudes in the northern hemisphere. Both the Humboldt Current along the Pacific coast of Chile and the current of the Malvinas along the Atlantic coast of Argentina are cold, but the coldest air comes off Antarctica. Should this air happen to cross a greater stretch of open ocean before crossing over South America, temperatures are moderated, but if

Windblown willows and lenticular clouds near Sarmiento, Chubut.

the air travels along the Antarctic Peninsula before crossing over the ocean, it is the coldest air mass possible.

The coldest temperatures experienced in Patagonia are not at its far southern tip but in its central region: Correa et al. (1969–98) mentions a low of −25.6°C in Maquinchao, Río Negro, with a close second of −24.2°C in Paso de Indios, Chubut. The official record low for Argentina was in Sarmiento, Chubut, where it reached −32.8°C in 1907.

The annual average temperature of the steppe itself is usually below 13°C. Extended periods of summer heat are rare. Though temperatures above 40°C have been recorded from Río Negro and Neuquén, the far southern reaches of Santa Cruz rarely go above 27°C. Temperature swings between day and night can be great, however, with hot days almost always followed by cool nights.

PRECIPITATION

Temperature and precipitation define climate and the resulting vegetation. The steppe of Patagonia shares a semi-arid climate with all steppe climates worldwide. Its precipitation regime in the western and central parts is similar to that of the Great Basin and steppes of eastern Washington, Oregon, and British Columbia; 46 percent of the precipitation falls in the winter, while only 11 percent falls in the summer (Paruelo et al. 1998). Closer to the southern Atlantic and its weather patterns, precipitation is more evenly distributed through the year, and some of the aridity is moderated by ocean humidity. But even on the Valdes Peninsula, for

example, precipitation is concentrated between May and October, the southern hemisphere winter.

WIND

The anticyclones of the South Pacific and Atlantic create a surprising pressure gradient, with resulting high winds over the steppe; indeed, blasting winds, particularly through the warmer months, are what capture the attention of most visitors to Patagonia. In addition to literally shaping the flora, these winds help scatter volcanic ash from the Andes across the steppe and eventually into the South Atlantic. High wind velocities during the growing season increase drought pressures on the vegetation by raising evaporation rates.

SOILS

Soils on the steppe vary, depending on the local geology and history of the area. Basaltic flows have covered the sedimentary rocks in some spots, but where the sedimentary rocks are not buried, soils are clayey or sandy. Large ash falls impact the soils in some places, and in other areas fluvial-glacial outwashes create the growing media for plants. Most soils tend to be well drained and low in organic matter.

Habitats and Plants

The steppe and surrounding regions have been divided into 16 floristic units; two are in the Monte province, two in ecotones where the Monte and steppe meet, and the rest fall in the six districts of the Patagonian phytogeographic province: Western, Central, Gulf of San Jorge, Subandean, Magellanic (or Fuegian), and Payunia (León et al. 1998).

WESTERN

This district, west of the 70th meridian, forms the western boundary of the steppe. The boundary is imprecise and forms a large ecotone in itself, stretching from the mountains between Loncopué and Chos Malal in Neuquén to Lago Buenos Aires in Santa Cruz, and east onto the sierras and tablelands of Río Negro and northwest Chubut. It is characterized by grasses and shrubs 60 to 180 cm tall, with nearly 50 percent vegetation coverage.

CENTRAL

This is the most extensive of the six districts. Located from the northwest of Maquinchao, Río Negro, to Río Coyle, Santa Cruz, it is more arid than the Western District and areas to the east, with precipitation of less than 200 millimeters

The Western District, with its greater precipitation, is grassier than the Central District.

per year. The region is characterized by having vegetative cover of no more than 50 percent, made up of one of the following communities, defined by particular shrubs: high shrub steppe dominated by *Colliguaja integerrima*; low shrub steppe by *Mulguraea tridens*; or, in central Chubut, steppe dominated by *Chuquiraga aurea*, *C. avellanedae*, and two species of *Nassauvia*. Another plant community, not so widespread, is dominated by *Nardophyllum*. All these shrubs have some ornamental use in landscapes within Patagonia and possibly beyond. Additionally, shrubs and herbaceous vegetation mix in with the various defining shrubs, depending on the location.

GULF OF SAN JORGE

This district is associated with the high plains that encircle the Gulf of San Jorge, from Cabo Raso to Punta Casamayor. It can be further divided into either high shrub steppe or grassy shrub steppe. Some floristic elements of the Monte province (*Prosopis denudans*, *Larrea ameghinoi*) have their southern limits here.

Nothofagus forest hugs the hillsides and lower mountain slopes above the Subandean District, signifying the start of the subantarctic province near El Chaltén.

The Payunia District in northern Neuquén has elements of the Monte and steppe. *Senna* and *Maihuenia* are typical plants of the region.

SUBANDEAN

This grassy district, a narrow strip from north to south between 71° and 71° 31'S, marks the meeting of semi-arid Patagonian steppe with the subantarctic province. Found where annual precipitation exceeds 300 millimeters, it forms a large ecotone with the nothofagus forest and in many places is a veritable mosaic of vegetation. This forest steppe ecotone is continuous only between 43° 30'(Tecka) and 46°S (Río Guenguel).

MAGELLANIC

This district is south of Santa Cruz province and north of Tierra del Fuego, between 52° and 54°S. It consists of two types of vegetation: xeric grassy steppe in the driest areas, where the vegetation cover is between 50 and 80 percent, and humid grassy steppe, found in areas with more rainfall.

PAYUNIA

This, the sixth district of the Patagonian phytogeographic province, encompasses the vegetation of southern Mendoza and northern Neuquén. Here the steppe reaches its highest elevations, occurring even above 1800 meters. At lower elevations, the steppe intergrades into the Monte province.

Larrea divaricata dominates the scene near Valcheta, Río Negro.

RIGHT Floristically, the Río Negro ecotone is intermediate between typical Monte and the Patagonian steppe.

MONTE

Higher temperatures and lower precipitation create Monte, the shrub-based ecosystem that replaces steppe when the average temperature exceeds 13°C. Monte is found north and east of the steppe proper. Typical (southern) Monte is characterized by *Larrea*, *Prosopis*, and several other genera also found in the deserts of southwestern North America. In eastern Monte, northeast of Río Negro, annual precipitation is greater than 250 millimeters; it has a greater vegetation cover (50 to 80 percent) than typical Monte.

ECOTONES

The transition between the Patagonian phytogeographic province and its neighbors is influenced by air masses from the Atlantic Ocean, which cause a gradual change in both temperature and precipitation. The shift is further complicated by changes in elevation. Two ecotones occur where the Monte and steppe meet: the Valdes Peninsula ecotone (including the Ameghino Isthmus) has salt flats in its low-lying areas; the Río Negro ecotone occurs mainly on slopes 300 to 600 meters above sea level.

WETLANDS

All the various ecotones are punctuated with river, lakebed, and other wetland habitats, supplied with either fresh- or saltwater. These habitats don't cover large areas in the steppe, but together their importance to wildlife and influence on vegetation are paramount.

Malline is a Spanish word used to describe any moist area of the steppe or alpine environments of southern South America; such an area often has an accumulation of organic matter similar to the sedge peat found in steppe fens of

A typical spring-fed malline with *Cortaderia araucana* and exotic *Salix ×fragilis*.

western North America. On the steppe, mallines, whether seeps, springs, or moist meadows, are most often low-lying sedge- (*Carex* spp.) and grass-dominated areas—very different, floristically, from the surrounding steppe. They are most common in the Western District.

Riparian areas line the many rivers and arroyos (streams) throughout the steppe. In general as one moves east, the number of perennial arroyos arising on the steppe itself decreases; and with the decrease in precipitation in the central and eastern part of the steppe, many arroyos are merely seasonal. Only the large rivers originating in the Andes (and the odd arroyo supplied by large springs) continue to flow through the summer. The vegetation along these rivers and arroyos shifts with the nature of the water; some plants are tolerant of saline conditions, others are not.

Halophytic vegetation is an important component of all steppes, with their dry salty lakebeds. Patagonia has some halophytic genera in common with other

Salicornia is found on saline soils throughout the world's steppes.

RIGHT *Anarthrophyllum strigulipetalum* showing the evolutionary response to grazing pressure: each leaf is tipped with a hard sharp spine.

steppe regions, including *Atriplex* and *Salicornia* (glasswort); its halophytic plants are most frequent in the dry lakebeds of the Central District.

EVOLUTIONARY FACTORS

One cannot visit the steppe of Patagonia for long without noticing the high percentage of spiny plants. Spination is usually an evolutionary trait developed by plants as a defense against grazing animals. All *Anarthrophyllum* species and almost all mulinums—as well as many adesmias, various junellias, and certain members of Asteraceae, Solanaceae, and other families—have spines or spiny leaves to deter grazing. Some debate surrounds this spiny flora: traditionally it was thought there was not much grazing pressure in Patagonia, but Lauenroth (1998) has suggested that guanacos were once plentiful enough to supply it. Moreover, many more species of herbivore megafauna, most famously the ground sloth (*Megatherium*), became extinct in the late Pleistocene; the impact these herbivores might have had is hard to assess.

Perhaps even more noticeable than the percentage of spiny flora is the percentage of cushion-forming species, but the reason this time is obvious: wind = cushions. Plants often take on cushion, bun, or mound shapes as an adaptation to average high wind velocities. Alpine, arctic, steppe, and seaside environments—all of which experience high rates of wind—have cushions as part of the flora, and all steppe regions worldwide have their particular cushion-forming species: Central Asia has *Acantholimon*, and western North America *Astragalus*, *Eriogonum*, and *Phlox*, for example. Patagonia is home to many genera of especially large-sized cushions, some specimens approaching one or more meters in diameter; among the largest cushion-forming species are *Chuquiraga aurea* and *Petunia patagonica*. Inspection of damaged cushions often reveals large woody branches

Petunia patagonica cushions on the Santa Cruz steppe.

the size of a man's wrist: presumably such specimens are of an age commensurate with their large size.

Touring the Patagonian Steppe

Patagonia may resemble other steppe regions in geographic features, but its flora is at once strikingly similar and vastly different. Takhtajan (1986) explains the situation concisely: "Through the Andean chain [...] the flora of Patagonia is connected with the temperate floras of the northern hemisphere." And yet? Though its diverse flora contains "a considerable admixture of taxa of Holarctic origin," it is crucially and basically derived "of ancient Holantarctic elements." The "transtropical high mountain bridge" that is the Andes allows for some species to inhabit both the North American and South American steppes, and in a few examples (thanks to the Bering Land Bridge) even the Asiatic steppe. *Gentiana prostrata* is found in three of the four steppe regions. While there are many similarities between the steppes, Patagonia, like all steppe regions, is also a center of endemism, including (besides the aforementioned *Benthamiella*) the monotypic *Ameghinoa*.

In both North American and Eurasian steppes, *Populus* species and other trees inhabit watercourses and other favorable habitats, but one striking aspect of the native steppe flora in South America is the almost complete absence of trees,

Ameghinoa patagonica.

Gentiana prostrata in the High Andean province with *Anagallis alternifolia*.

RIGHT *Salix humboldtiana* growing along the Río Chubut, its southernmost location.

even along waterways. Only one species of willow is native in the central and northern part of the steppe: *Salix humboldtiana* reaches its southernmost limits along the Río Chubut near the town of Las Plumas, north of the 44th parallel. Many small to large shrubs find suitable habitat within the steppe, and in certain areas shrubs from various families (Asteraceae, Anacardiaceae, Euphorbiaceae, Amaranthaceae, Fabaceae, Solanaceae, Verbenaceae) tend to dominate the landscape. These dominant families are common to all steppe regions, with the exception of Solanaceae and Verbenaceae, which are uniquely diversified in Patagonia, filling a number of niches. Other ways in which the Patagonian steppe is set apart from (or tied back to) its sisters in the northern hemisphere can be observed as one travels through the steppe.

Starting in northern Patagonia, we will take a counterclockwise tour, moving first south along the spine and base of the Argentine Andes; then heading east from El Chaltén (leaving the Andean peaks slowly receding in the rearview mirror) to the wild and windy coast near Comodoro Rivadavia and points north; and finally back across again, to the Andean foothills.

From Neuquén to El Chaltén the Patagonian steppe abuts the Andes in a continuous belt that varies in width depending on the latitude. In the north, the steppe is narrow, hugging the Andes from Río Negro to Chubut; it gets increasingly wider, until it spans the distance from the Andes to the Atlantic in southern Chubut, Santa Cruz, and Tierra del Fuego.

PRIMEROS PINOS TO DINA HUAPI

Although it is north of the 39°S latitude line that defines Patagonia, Primeros Pinos is near the point where the steppe begins to hug the base of the Andes. Continuing north to at least Mendoza, the steppe clings to the base of the Andes and in places climbs the slopes, replacing the nothofagus forest north of 38°S. Eastern slopes of the Andes north of 38°S are largely treeless, with no nothofagus or other arboreal vegetation to replace them. This is likely attributable to the more severe rain shadow created by the increasing altitude of the Andes. Primeros Pinos is named for the stands of *Araucaria araucana* that straddle the transition zone from steppe to the Andean forest proper of *Nothofagus* and *Araucaria*.

Driving roughly southwest from the city of Zapala, one transitions from the Monte into the steppe proper. The steppe slowly rises until finally snow-studded mountains appear above it. This range, sparsely covered in a patchy araucaria-nothofagus forest, is an eastern extension of the main Andes chain, and the araucarias are the easternmost representatives of their tribe; here they cling precariously—islands of trees in an otherwise treeless landscape. The first grove is stunted and tortured, hugging a slope with steppe vegetation at its base. A few other scattered groves dot the surrounding mountainside (as ponderosa pine groves do the rocky hillsides and escarpments of the western Great Plains). An alpine pass separates these araucarias from their lowland brethren on the other side.

As one climbs, it is easy to see the relationship between alpine and steppe floras. This pattern of seamless melding repeats itself around the globe, wherever steppe and alpine rub shoulders in rain shadow areas. Many species found on the steppe grow next to species generally found only in the alpine. *Grindelia prunelloides* has a particularly showy and floriferous form—an adaptation of many alpines to attract pollinators in a colder and even windier climate than the steppe proper. *Viola dasyphylla* grows with various nassauvias. The transition from alpine proper to forest is especially abrupt here; no krummholz marks the timberline. Instead, sizable araucarias rise up from the rocky landscape, and stunted groves of nothofagus hug the moister slopes.

In the distance, on the spine of the Andes, are the stratovolcanic cones of Lonquimay and Llaima, an occasional wisp of smoke rising from their calderas. An increase in precipitation on this side of the pass allows more continuous araucaria forest, yet still sparser than groves in far western Argentina or Chile, and there Río Alumine gathers its waters from the surrounding mountainside and begins its journey southward to join the Río Lamy on the steppes of Neuquén. At lower elevations, the araucarias give way to plantations of *Pinus ponderosa*; North American pine species are heavily planted in Argentina for timber production and forestation projects.

Viola dasyphylla.

Araucaria araucana at timberline near Primeros Pinos.

Likewise descending in elevation, the Río Alumine enters into a canyon that could fit visually into western North America or Central Asia with little modification. The almost perfect cone of Lanín appears up a side valley as the road leaves the river valley and snakes toward Junín de los Andes. Junín hosts a giant gaucho festival each summer, the South American gaucho being the equivalent of the North American cowboy. The heat, dust (ubiquitous in all steppe summers), and excitement recall summer rodeos in any western North American town, from Alberta to Chihuahua.

Lanín is hidden from Junín by a ridge, but a short journey toward Lago Huechulafquen brings the volcanic peak into view. A grove of araucarias set amid steppe vegetation frames the first glimpse of blue Lago Heuchulafquen. In a pattern repeated throughout Patagonia, the steppe passes through an ecotone as rainfall increases from east to west. By mid-lake, a stately forest of *Nothofagus dombeyi* reigns supreme, and steppe—even an occasional pocket meadow of needle grass—clings to sun-baked rocky outcrops.

From Junín to San Martín de Los Andes, the road passes through a valley similar to many valleys west of the Continental Divide in North America. Windbreaks of ponderosa pine and Rocky Mountain juniper make this area almost indistinguishable from its North American cousin—until one observes the araucaria-nothofagus forest in the distance.

Dina Huapi sits on the steppe at the east end of Lago Nahuel Huapi. Here the Río Lamy spills from the glacial basin of Lago Nahuel Huapi and begins a winding journey through the volcanic foothills of the Andes and steppe toward its rendezvous with the Río Neuquén coming from the Andes to the northwest. From Dina Huapi it is possible on clear days to look west to the spine of the Andes, clothed in dense forest that slowly peters out to the steppe surrounding Dina Huapi itself. A 45-minute drive brings the traveler to the base of Cerro Catedral.

CERRO CATEDRAL

Cerro Catedral ("cathedral peak") is well known to alpine plant enthusiasts for its rich and diverse alpine flora and ease of access, courtesy of the chair lift to its summit. The chair lift can be full of tourists from warmer climes, hoping to touch snow and throw snowballs in midsummer—a spectacle that repeats itself in the ski resorts of western North America. While it would appear that the top of an alpine mountain would have little in common with steppe thousands of feet below, most genera and even many of the same species of the High Andean province are also found on the Patagonian steppe, on dry rocky outcrops in the mid-elevations and again above treeline—just as alpine taxa recur in steppes below the Western Cordillera of North America, the high peaks of Central Asia, and the mountains of South Africa.

Viola sacculus.

Loasa nana.

Furthermore Cerro Catedral illustrates an interesting phenomenon found elsewhere in the southern Andes and mirrored in the Cascades of western North America: the more arid the high mountain, the greater the diversity of flora. A transect set up in Nahuel Huapi National Park between three mountains illustrates the point, with the easternmost peak being the most diverse and the westernmost, the least. It is the same on the east and west sides of Mount Rainier: Paradise on the wetter west side is somewhat species-poor due to the long season of snow cover, while Sunrise on the drier east side has species-rich meadows that do not exist at Paradise.

The area around the Cerro Catedral chair lift appears bleak, at first. *Nassauvia revoluta*, looking like a plant from Mars, lets visitors know they are not in the northern hemisphere anymore. *Oxalis adenophylla* peeks out from between rocks, kept company by various hardy senecios that would look right at home in Colorado scree. Classic Andean flora hug the ground along service roads, trying to hide from the wind: *Viola sacculus* and *V. cotyledon* (both rosulate violets), *Mulinum leptactanthum*, *Loasa nana*, and that queen of Patagonian snowbanks, *Ranunculus semiverticillatus*. Streamsides are home to yellow *R. peduncularis* and screaming-red *Ourisia ruellioides*. Scree and rocky slopes hold *O. fragrans* and *O. alpina*.

At the bottom of the mountain, forest steppe ecotone replaces the thick nothofagus forest. *Embothrium coccineum* blooms in spring, painting heart-stopping patches of red on the distant hills. Roadsides are a mix of species, of which *Alstroemeria aurea* steals the show in midsummer, with unusually large flowers of bright yellow or saturated orange. The forest steppe ecotone and adjacent steppe would prove to be the most interesting for a wide array of species with horticultural application, as the plant primer will show.

Nassauvia revoluta.

TOURING THE PATAGONIAN STEPPE

Rhodophiala mendocina on the steppe east of Bariloche.

RIGHT *Mutisia spinosa* on a fence near Esquel.

EAST OF BARILOCHE AND ESQUEL

As even the forest steppe ecotone fades into the expanses east of the Andes, large woody plants slowly disappear and a mix of grasses, herbaceous plants, and small to medium shrubs take over. Midsummer on the steppe can be colorful, yet because most precipitation falls in the winter, many things are done flowering or even dormant by mid-January.

In the open fields left between houses in eastern Bariloche, amazing diversity can be found: bright orange *Mutisia decurrens* ornaments not only moist nothofagus forest and sunny forest steppe ecotone (often crawling over *Mulinum spinosum*), but also candy tins, hand-painted plates, and undeveloped lots as well. Moving out onto the steppe proper east of the airport, midsummer brings on *Rhodophiala mendocina* and *Tropaeolum incisum*, both steppe extensions of mainly Chilean genera. The rhodophiala is a relative of the common amaryllis (*Hippeastrum* spp.) but with smaller trumpets of soft yellow. The tropaeolum steals the show with its peachy apricot flowers (pale yellow in a few individuals). *Mutisia spinosa* is pretty in pink, on low rocky hills where *Austrocedrus chilensis* makes its final stand before submitting to the increasing dryness of the steppe. Further down the road, *Junellia succulentifolia* brightens a roadside bank, and the charmingly acid yellow-green of *Mulinum spinosum* contrasts with the predominant tans and browns. Near the bridge on the Río Pichileufú, *Austrocactus patagonicus* perches on a rocky cliff. Large mounds of *Junellia tonini* cover themselves in a mix of lavender and rose flowers each spring (mid-November) into summer.

South of El Bolsón, famed Route 40 swings east. Isolated specimens of *Cortaderia araucana* pierce moist seeps. The Andes continue to hold the western horizon, forming a visual boundary for the steppe. Emerging onto the steppe

Tropaeolum incisum.

from the Andes is always breathtaking, as forest and foothills give way to sweeping views of distant earth-colored sierras. In places, one of the large rivers bursts forth, seeking the Atlantic, but more often than not the attempt is futile and the exit is waterless, a pattern of steppe climates: it is the shortage of water that creates this environment.

EAST OF CORCOVADO NEAR GOBERNADOR COSTA

East of Corcovado the transition from nothofagus forest to steppe is dramatic. Pale green cushions of *Azorella patagonica* occupy a dry lakebed, creating not only a color but also a textural contrast to the scruffy landscape, an open area dominated by the coffee tones of *Nassauvia axillaris* and *N. glomerulosa*. The various nassauvias mimic the sagebrushes of the North American steppes, and the giant cushions of the azorella, 1 to 1.5 meters across, are quite a sight. On the margins of the dry lake, a small yellow composite, *Grindelia prunelloides*, can be found. Near Tecka, specimens of *Maihuenia patagonica* stand out, bright green in a sea of tan and buff; this cactus species is found from Mendoza to Santa Cruz, thriving in both the steppe and Monte. North of Gobernador Costa, the ever-increasing presence of cushions is felt: the bright yellow-orange flowers of *Brachyclados caespitosus* are a startling contrast to the surrounding drab landscape, and amazingly tight neon-green azorellas and a silvery adesmia often create "bunneries"—along with broken liquor bottles and the bleached bones of livestock, a scenario all too familiar in the steppes of Wyoming, Nevada, or western Colorado. A dominant cushion of the drier parts of the steppes, from Chubut to Santa Cruz, *Chuquiraga aurea* grows beside the road from Tecka to Gaiman; large spines with golden strawflowers dot its cushions.

WEST OF RÍO MAYO TO LOS ANTIGUOS

Río Mayo, like many towns on the steppe, sits in a sheltered site at the base of some bluffs. It is too far from the Andes to offer any comforting glimpse of snow-covered peaks, and the overwhelming emptiness of the plain can be felt in both the incessant wind and lack of geographic features. The flora expresses the characteristics of the Central District. Large cushions of brachyclados and azorellas, low shrubby adesmias, and *Stipa humilis* fill out the picture. Moving southwest toward the border of Santa Cruz and eventually Perito Moreno, that vegetation continues with various species—junellias, *Acaena caespitosa*—filtering in or out, depending on minor differences in the flat topography. Midsummer brings the last fairy slipper flowers on the green rosettes of *Calceolaria polyrhiza*, trying its best to make up for the featureless plain upon which it grows. All are adapted to bloom in the spring and early summer, when soils still hold moisture from winter rain and snow. From here on out, even an occasional rainstorm is not enough to bring on any kind of summer display, and the plants hurry to set seed before winter cold returns. Such is the pattern of the winter-precipitation steppes around the world: a brief spring show that can border on spectacular in kind years, followed by the tenacious flowering of tap-rooted species that use water deep in the ground to feed nectar-seeking pollinators and brighten the journey of the equally tenacious traveler, coated in dust and parched by wind and summer heat. The road slips quietly and without ceremony across the Chubut and Santa Cruz provincial boundary, as many a dirt road in the North American Great Plains does. Nature pays no attention to arbitrary borders. The same vegetation blankets the landscape, undisturbed.

The plain finally ends at some low hills, the rim of the basin in which Lago Buenos Aires sits. Here the abrupt change in topography creates some new habitats for the ever-savvy plant life of the steppes. In rocky outcrops near the lakeshore, *Colliguaja integerrima*, a shrubby euphorbia relative with evergreen leaves,

A roadside shrine near the town of Perito Moreno. *Lycium chilense* grows on the sheltering rock outcrop.

forms large colonies, as it does in favorable locations from the base of the Andes to the Atlantic. Perito Moreno and Los Antiguos nestle in somewhat sheltered locations, the climate of the latter so mild that sweet cherries and soft fruits are cultivated in the rich volcanic soils.

LA MESETA DEL LAGO BUENOS AIRES

High plateaus are found along much of the length of the Patagonian Andes. Most are volcanic in origin, and Meseta del Lago Buenos Aires too is speckled with cinder cones and an occasional crater. At 900 to 1100 meters on average, this immense plateau—too dry and possibly too high for tree growth—is covered with high-elevation steppe flora. Various *Benthamiella* species and *Anarthrophyllum desideratum* are dangerously close to their northern limits here. Sundry azorellas, *Satureja darwinii*, various cushion-forming shrubby senecios, *Nassauvia glomerulosa*, and an assortment of other species try their best to cover some ground just above plateau's rim. An area where volcanic ash from the 1991 eruption of Cerro Hudson accumulated among giant bunchgrasses has created a striking landscape of hummocks fit for a giant. The flat purple cushions of *Junellia micrantha* pool in a dry lakebed in the equivalent of a southern hemisphere alpine summer. Near a shallow lake, various genera (*Acaena, Nassauvia*) take the

Meseta del Lago
Buenos Aires with
Junellia micrantha.

evolution of cushion plants to a level of tightness experienced in few parts of the world. Herds of guanacos graze and roam about the larger-than-life landscape, watched over by Mount Zeballos, a volcanic peak rising high above the plateau to the west and worn by the endless passage of time.

MOUNT ZEBALLOS

Between Los Antiguos and Lago Posadas, a tortuous dirt road leads up over a pass between Mount Zeballos and the Chilean border. Winding up a series of switchbacks through thickets of *Colliguaja integerrima*, one eventually reaches a higher valley that could just as well be in the mountains of Nevada or eastern Oregon. The road climbs again up the side of the valley, passing through sparse nothofagus forests; a similar forest can be found in a few rocky locations where

The steppe near
Mount Zeballos.

LEFT *Astragalus
cruckshanksii*.

quaking aspen (*Populus tremuloides*) or ponderosa pine (*Pinus ponderosa*) meet, blending from open forest steppe to steppe. At last, beyond all trees, the landscape transforms into a rich green grassland (*Festuca* spp., *Mulinum spinosum*) and grazing cows. Higher yet the meadows give way to open scree slopes and patches of barren earth; *Nassauvia pulcherrima* colonized these areas. Mallines are dominated by giant cushions of *Azorella lycopodioides* and mats of sedges and small grasses. *Gentianella magellanica* finds a suitable home in these moist areas, as does *Gentiana prostrata*, which is also found in western North America and Central Asia. In areas not occupied by screes or mallines, mats of *Empetrum rubrum* (related to the northern hemisphere's *E. nigrum*) and other low-growing taxa clothe the slopes. In openings, *Viola auricolor* (the most southern of the rosulate violets) and *Astragalus cruckshanksii* make an appearance. The astragalus would be right at home in western North American or Central Asia. The very top of the pass is a barren and quiet world of rock and scree, the only sound besides the wind being the occasional screech of a gliding raptor, a pattern repeated on any high summit around the world. *Azorella ameghinoi*, *A. monantha*, *Valeriana moyanoi*, various senecios, nassauvias, and erigerons—all hunker down between rocks, seeking shelter from the drying winds and sun. The presence of *Viola*, *Astragalus*, *Gentianella*, *Empetrum*, *Erigeron*, *Senecio*, and *Valeriana* shows the relationship of the Patagonian flora with that of the northern temperate zones and especially that of steppe regions.

The various vegetation zones are repeated down the other side of the pass, except for the nothofagus forest. Instead, the alpine merges into montane steppe and eventually into dry steppe, dominated by *Mulinum spinosum* and *Colliguaja integerrima*. The dry lakebeds and rock slopes at lower elevations are home to their own treasures. Bright green mats of *Lycium repens* stand out against the pale dry lakebeds, and fuzzy cushions of an *Azorella* species look silver in the afternoon light, thanks to their fuzzy leaves. A collection of volcanic and nonvolcanic peaks hems in the western end of the valley. While these are awe-inspiring, nothing prepares one for the stark contrast of blue-green Lago Posadas and darker blue Lago Pueyrredón, separated by a narrow spit of land and draining—against great odds—toward the Pacific through the crest of the Andes.

LAGO POSADAS TO EL CHALTÉN

The small town of Lago Posadas sits to the east of its namesake lake but still within its drainage basin, and east of town, the bright green of *Lycium repens* once again dots dry lakebeds joined by the larger *L. chilense* and an atriplex (possibly *Atriplex lampa*). Similar communities, dominated by the same families and often the same genera, can be found in western North America. Once out of the basin, the undulating steppe takes over, the broom-like cushions of *Fabiana nana*, resembling a pack of hedgehogs, scattered across its dry brown slopes. This species demonstrates what the nightshade family has done in South America: solanaceous plants have here become medium to large cushion- or mat-forming woody shrubs. Now-silvery mounds of *Brachyclados caespitosus* mix with the various blond needle grasses (especially *Stipa humilis*), painting a familiar picture: silver and blond are defining colors of all steppe regions in the dryness of late summer.

Near Bajo Caracoles, Route 40 begins to climb a series of wide benches, topping out at 900 meters above sea level. Junellias and a few other shrubs offer a modest break from the largely brown grasses and sleeping cushion plants; and the very highest points are home to various cushion- and mat-forming communities, with benthamiellas and ephedras, among others, fighting for stability on these wind-blasted hills. Farther south on the plains are small mesas formed from basaltic lava; large areas of these are dominated by *Mulguraea tridens*, a woody verbena relative with leaves so dark that, from a distance, plants appear black and dead. In this land of black rock, black shrubs, and blond grasses, *Polygala darwiniana*, *Silene antarctica*, *Anarthrophyllum desideratum*, a wickedly spined adesmia, and a cushion-forming acaena gain a foothold. Both *Anarthrophyllum* and *Adesmia* take a variety of forms in Patagonia, from low mats and cushions to large spiny shrubs.

South of the basaltic mesas, the land reverts to outwashed plains, rolling hills, and bluffs that slowly melt into plains of the same material at their base.

Fabiana nana on the steppe east of Lago Posadas.

The silvery cushions of *Adesmia suffocata* near Tres Lagos.

LEFT On the road to El Chaltén, a wind-blasted junellia, with Lago Viedma in the distance.

Everywhere are shades of tan, buff, brown, and blond. Only hardy green cushion plants break the monotony of earth tones. Cushions of *Chuquiraga aurea* dot the clayey roadcuts, soon followed by refreshingly bright and amazingly tight mounds of *Petunia patagonica*, eking out an existence on gravelly outwashes. An abnormality in a genus of what most people think of as blowsy annuals, this long-lived woody plant forms sizable mounds 2 to 3 meters across and more than half a meter high; traditionally shy-flowering in cultivation, a pitstop survey of the population revealed both heavily flowering individuals and individuals that had no spent blooms.

Route 40 rolls southward, oblivious to the bright green cushion along its margins. Near the hamlet of Tres Lagos, it crosses a series of hills and bluffs, open to the full blast of the westerly wind. Here a particularly nice sample of the cushions can be found. *Adesmia suffocata* shares the rocky hillside with *Brachyclados caespitosus* and *Frankenia patagonica*, and as the milky blue glacial waters of Lago Viedma come into view, the landscape holds *Mulinum spinosum*, *M. echinus*, gray *Senecio filaginoides*, and bright green *Nardophyllum chiliotrichoides*.

Like many towns in Patagonia, El Chaltén sits right in the middle of the forest steppe ecotone. *Embothrium coccineum* crops up among rocks, while various needle grasses and *Anarthrophyllum desideratum* pick even more xeric spots. Greater precipitation allows the nothofagus forest to inch down from the slopes of the Andes, which here include the skyscraping spires of Cerro Fitz Roy and

Anarthrophyllum rigidum, a taller shrubby anarthrophyllum, can resemble certain Eurasian peashrubs (*Caragana* spp.).

neighboring Cerro Torre, creating a landscape that blends the green of various plants not tough enough for the steppe with the steppe's tan and blond.

PUERTO SAN JULIÁN TO BOSQUES PETRIFICADOS

And now to head east from El Chaltén, to the coast. Little breaks the undulating and generally overgrazed steppe between Tres Lagos and Puerto Santa Cruz. Scattered ranches in the distance with their windbreaks of Lombardy poplars; the abandoned wide spot in the road, where the town of Los Olmos once stood, with its namesake Siberian elms from the steppes of Central Asia; and ubiquitous wire fences—these are the only signs of the human dream to settle the steppe. Once on the major coastal highway, the Gran Bajo de San Julián appears to the west—at 105 meters below sea level, the lowest point on the South American continent. Old dry lakebeds provide some of the lowest points on three continents: the Gran Bajo de San Julián, Death Valley, the Junggar Basin in China—all are in steppe or desert regions. Any sign of green is welcome by this point in the journey through a treeless world, and north of Puerto Santa Cruz, *Colliguaja integerrima* and *Anarthrophyllum rigidum* brighten the hilly landscape with just that.

Off the main coastal route toward the Monumento Natural Los Bosques Petrificados ("petrified forests"), *Chuquiraga aurea* enlivens rocky hummocks

Ameghinoa patagonica,
close-up of leaves.

LEFT *Prosopis denudans.*

and hillsides, along with limoniums and a glasswort (*Limonium* is found in South Africa and Eurasia; *Salicornia* is found in all steppe climates). Closer to the natural monument, a series of rocky ridges and arroyos could be anywhere in the U.S. Southwest: several species of *Prosopis*, *Larrea*, and *Lycium* flourish here, further strengthening the relationship between Patagonia and western North America. Early spring on these rocky outcrops must be as colorful as it is west of the Continental Divide in North America.

Indeed, the landscape around Bosques Petrificados could easily be the Painted Desert of northeastern Arizona, with substrates running from almost white to reds, oranges, burnt umbers, tans, browns, and grays; *Prosopis denudans* (with its bright red seed pods), various small cacti (*Maihuenia patagonica*, *Maihueniopsis darwinii*, *Austrocactus* spp.), and a spiny adesmia (trying its best to mimic a cactus) further recall the look of that U.S. state. *Gutierrezia baccharoides*, a small asteraceous subshrub, stakes its claim in this rocky world, as *G. sarothrae* (broom snakeweed) does in the North American shortgrass prairie. An eroded volcanic plug, Madre y Hija ("mother and daughter"), dominates the distance. Vegetation is sparse in general, but azorellas add sprigs of greenery, and *Frankenia patagonica* and *Ameghinoa patagonica* eke out a living in the openings between logs and stones, trying their best to prove that not all life died when the proto-araucaria forest was buried in ash millions of years ago.

Ephedra frustillata can form impressive cushions.

RIGHT *Pleurophora patagonica*.

LAS HERAS TO SARMIENTO

The flora becomes richer as one climbs the series of rolling hills west of the oil boom town of Las Heras. Various species of *Schinus*, *Mulguraea ligustrina*, *Chuquiraga avellanedae*, and *Ephedra ochreata* are just a few of the woody shrubs whose greenery is welcome in the tan and treeless landscape, as are the rose-pink flowers of *Pleurophora patagonica*, one of the few summer-flowering plants in the drier parts of the steppe. On hillsides full of fossilized clams, *Stipa humilis*, *Anarthrophyllum desideratum*, and several species of *Festuca* and *Sisyrinchium* keep an inconspicuous (small, brownish) pterocactus company. Mat- to cushion-forming *Ephedra frustillata* produces fruit-like structures that are white, not the common red or orange; cushions of it dressed one wind-scoured hilltop just before the descent into the great lake basin near Sarmiento. Sarmiento lies in the verdant valley that holds Lago Musters and Lago Colhué Huapi, both fed by the Río Chico—a true oasis in a largely brown and sparsely inhabited part of Argentina.

BUEN PASTO

North of Sarmiento, both altitude and precipitation increase. The transition from the desert-like vegetation surrounding Sarmiento to the more grass-dominated steppe by Buen Pasto ("good pasture") is marked by a quasi-chaparral of various shrubs, including *Fabiana patagonica*, *Pleurophora patagonica*, and *Maihuenia patagonica*. Vining through the community is the steppe endemic *Mutisia retrorsa*. Compared to other Patagonian mutisias, which are found in the more humid forest steppe ecotone, high mountains, or nothofagus forests, this species occupies much drier habitats; it occurs from Mendoza to Santa Cruz, but we saw it only here.

On the summit of one of the many small sierras of Chubut and Río Negro, some of the chaparral-like vegetation shifts to species found farther west on the steppe. *Mulguraea tridens* dominates rocky slopes, and *Azorella patagonica*, once again, a dry lakebed. *Ephedra frustillata* crawls over rocks, and there were the most beautiful gold-spined forms of *Austrocactus patagonicus* that we would see. North of Buen Pasto a hike to another hilltop revealed silver forms of *A. patagonicus*, large cushions of benthamiellas, and a bright green acaena. The wind on this particular summit was unbelievable—actually driving me back at times.

Austrocactus patagonicus, gold-spined form, near Buen Pasto.

COMODORO RIVADAVIA TO TRELEW

Sarmiento and Comodoro Rivadavia are their own floristic unit, part of the Gulf of San Jorge District. Once out of the Sarmiento's lake basin and heading for Comodoro Rivadavia, on the coast, one faces a broad flat plateau, with occasional breaks supporting massive patches of *Colliguaja integerrima*. Little else seems to grow in the landscape, but in areas truly dominated by this shrub, its bright evergreen leaves and texture are a welcome contrast to the general tans and browns. Comodoro Rivadavia's climate is mild, and a marked change in vegetation continues along the coast as one heads north toward Trelew. The Monte and steppe meet along a blurred line that runs from south of Trelew (44° 20′ S) all the way to Mendoza. *Larrea* becomes more plentiful on the Monte side of the line, where the average temperature exceeds 13°C. In Río Negro and Neuquén there exists a transition zone: steppe at elevations above 400 meters, Monte below.

TRELEW TO ESQUEL

This particular route, which follows the Río Chubut for much of its length, runs from the coast all the way to the forest steppe ecotone on the slopes of the Andes surrounding Esquel, passing through Monte and various steppe communities along the way. The endemic *Larrea ameghinoi*, the southernmost representative of its genus, was driven by Patagonian westerlies to its present creeping mat form; it is found in several stops between Trelew and Paso de Indios (a town that could serve as the setting for an old-fashioned western with its reddish, orange, and chocolate-colored landscape, dominated by *Larrea*). As the river valley slowly rises in elevation and the traveler moves farther from the ocean, the landscape becomes more and more steppe-like. By the time one reaches Pampa de Agnia, the Río Chubut valley, with its various canyons and farming communities, has been left behind, and a high plain dominated by *Maihuenia patagonica* stretches to the northern horizon. To the east the various sierras of Chubut form the skyline, and any traces of the Monte are left far behind.

From Pampa de Agnia to Esquel, the road rises and falls over various small sierras and ridges, in a pattern similar to the basin and range topography of

Arrowheads on the Meseta Somuncurá.

Nevada or Oregon, the only difference being the mountain ranges are not as dramatic. Near Esquel, the road passes through plantations of North American pines. The giant circular tour of the steppe is now complete, and it's fitting in many ways that *Pinus ponderosa* and *P. contorta* var. *latifolia*, two species of pines that brush the North American steppe and even infiltrate it in their native ranges, welcome visitors to Esquel after a hard day's drive across a treeless landscape.

Plant-People Connection

Patagonia is widely thought to be the last major part of a continent settled by humans; this occurred some 15,000 years ago (or perhaps even well before). Some of the earliest inhabitants left their astonishing artwork in the Cueva de las Manos and lesser-known sites. Descendants of these first inhabitants of Patagonia still reside in the region. The Mapuche are one such group, but like many of the nomadic peoples of the steppes worldwide, their lifestyle has been greatly impacted by European settlement.

Since Europeans first visited Patagonia, starting with Magellan in 1520, written accounts of the awe-inspiring grandeur of the land (and adventures had upon it) have filled many pages. Magellan went on to find his way through the strait that bears his name, and the *Victoria*, one of the original five ships in the expedition, became the first to circumnavigate the world. In 1767 Patagonia was again associated with a first: Jeanne Baret (just mentionably, the first woman to circumnavigate the globe, on the *Étoile*, under Louis-Antoine de Bougainville) is of special interest to botanists as she herself collected most specimens attributed to Philibert Commerson, especially along the Atlantic coast of Patagonia—the first herbarium specimens collected in Patagonia proper. Baret was trained as an herbalist and disguised as a sailor, and Commerson (her lover as well as the ship's botanist) depended on her to do most of the collecting while his health was in decline. For an interesting account of her life, see Ridley (2010).

After this expedition, the floodgates for scientific exploration of Patagonia were opened. Joseph Banks and Daniel Carl Solander passed through on their legendary trip around the globe, taking with them the first herbarium specimens collected in the Argentine part of Tierra del Fuego. Charles Darwin produced the first geological map of Patagonia around 1840 (now housed in the Cambridge University Library) and traveled to the steppe near Mendoza and on a journey up the Río Santa Cruz ("I always had a latent hope of meeting with some great change in the features of the country"); years later, in his autobiography, he wrote, "[T]he sense of sublimity, which the great deserts of Patagonia and the forest-clad mountains of Tierra del Fuego excited in me, has left an indelible impression on my mind." The observations Darwin made during his time here

An abandoned stone and plaster house in Neuquén.

helped him formulate his theory on natural selection and the origin of species.

The late 1800s were dominated by military campaigns. Ironically, as the Argentine government took up a campaign to remove the Mapuches from northern Patagonia, a similar drama was playing out on the steppes of western North America with the Great Plains tribes. As in North America the Mapuches were eventually defeated (the final battle was in 1884), and the land was opened to European settlement.

As much of Patagonia is suitable only for grazing, it was settled at a slightly later time than western North America; many towns date from the early 1900s. Patagonia's wide open spaces did not just spawn a cowboy culture (gauchos are Argentina's and Chile's version), it called to some of America's most notorious outlaws: Butch Cassidy and the Sundance Kid escaped to Patagonia and lived in Chubut from 1902 to 1907. Also in that province, a large influx of Welsh settlers established a thriving sheep industry, which by the 1920s had helped Argentina become one of the world's wealthiest countries; but the Great Depression struck there too, and over the last century, overgrazing, volcanic eruptions, and low prices on wool have led to a large-scale decline of that industry. Many former sheep farms are now turning to ecotourism, their abundant wildlife and spectacular scenery drawing visitors from around the world.

Patagonia is presently home to just under 2 million people residing across approximately 1 million square kilometers. Chile's residents tend to be squeezed between the sea coast and the towering peaks of the Andes; in Argentina, the major settlements are along the base of the mountains (much like the population centers of Nevada, Utah, and Colorado) or along major rivers and the coast of the Atlantic. The region is threatened by human activities, all of them driven by economics. The vast forest plantations of North American pines produce lumber, but several *Pinus* species and various other western conifers are naturalizing and may one day outcompete native trees and steppe vegetation, beyond the boundaries of the plantations. And in common with all steppe and grassland ecosystems worldwide, Patagonia has suffered from the vicious cycle of overgrazing and subsequent erosion. The final result is desertification, as once productive areas become barren and sparsely vegetated due to soil loss and the reduced ability of the remaining soils to hold water during dry periods.

Thankfully, a movement is afoot to encourage the cultivation of native steppe and southern Andean flora in the gardens of Patagonia. Important work is being done in Argentina itself by the Instituto Nacional Tecnología Agropecuaria (INTA) and by several private individuals in the Bariloche area. Many promising steppe species—*Corynabutilon bicolor*, *Nassauvia chubutensis*, *Senna arnottiana*, several senecios and junellias—are now cultivated in a handful of gardens in the region with good result.

The battle between the exploitation of natural resources and conservation rages on, and the opportunities for further research abound in all steppe areas. An especially rich area of exploration is the similarities between the two New World steppes. Leo Bruederle of the University of Colorado at Denver is conducting studies on *Carex* species common to Tierra del Fuego, the American Arctic, and the Southern Rockies. Another area of study would be the genera that appear to have disjunct distributions between North America and Patagonia (*Gutierrezia*, *Gaillardia*, *Eriogonum*, *Austrocactus*/*Sclerocactus*). The New World steppes may have been the last to feel the hand of man, but they are no safer than the one that first felt humanity's needs and desires—the Silk Road superhighway for conquest and cultural exchange in Eurasia. We have much to learn from these important regions before it is too late.

PATAGONIAN STEPPE

Alstroemeria aurea.

ALSTROEMERIACEAE

This family is endemic to the New World, from Central America to southern Patagonia; it consists of approximately 250 species in five genera, the most famous being *Alstroemeria*. Few flowers represent temperate South America in horticulture as well as alstroemerias; these natives of Chile and Argentina are well established throughout the northern hemisphere, from gardens in Mediterranean climates to bouquets in supermarkets. The hybrids involving *A. ligtu* are perhaps the best-known representatives of this large and varied genus. We will focus on two species that cross over the Andes into the steppe. The larger and better-known species, *A. aurea*, thrives in the forest steppe ecotone. True to its descriptor, its bright orange or glowing yellow flowers can paint entire roadsides shades of the sun. The large-flowered forms we saw on the east slope of the Andes surely have much value for cultivation in cold steppe climates. It is now listed as invasive in the mild Pacific Northwest.

Alstroemeria patagonica is a diminutive alstroemeria fit for a trough or rock garden. This denizen of the steppe seems to thrive in dry harsh areas. It has had a brush with horticulture, and several specialists probably cultivate it. I have killed it at least twice. What exactly it needs I have not been able to determine—perhaps wet, cold winters followed by cold, dry summers.

Alstroemeria patagonica.

APIACEAE

Carrots, celery, and parsley are three of the best-known members of a family found across much of the globe, from alpine summits to sandy subtropical seashores. Apiaceae is well represented in three of the world's major steppes (excluding only South Africa), but Patagonia more than any other is dominated by the plastic diversity of this large family, much of it driven by the high wind velocities that encourage plants to take on the shapes of cushions, mounds, and mats. Large cushions of *Azorella* and *Bolax* dominate many steppe communities, and throughout much of the western, central, and even eastern steppe, a closely related group of *Mulinum* species forms a visual leitmotif, often being among the largest plants around.

Mulinum spinosum, by the shores of Lago Pueyrredón and Lago Posadas.

Azorella monantha is one of the ubiquitous green cushions of the steppe and up into the alpine on the east slope of the Andes. This plant has only possibly flirted with cultivation. It would be a welcome addition to any rock, trough, or crevice garden.

Few plants in the world approach the cushion-forming ability of *Azorella patagonica*, whose nuclear greenish yellow cushions, up to a meter across, dot dry playas across the steppe. This plant might prove to be as amenable as *Junellia micrantha*, which has a foothold in cultivation and grows in similar situations (although we never saw them together in the wild).

Spiny cushion- to mat-forming *Mulinum* species have filled every niche on the steppe. *Mulinum spinosum*, one of the dominant and indicator plants of the Patagonian steppe, grows from high in the Andes to the coast; it has been cultivated on occasion and should be cultivated more. *Mulinum echinus*, a smaller

Azorella monantha.

Azorella monantha, close-up of cushion.

Mulinum echinus.

bluer version, is small enough for the rock garden, but true admirers of cushions and buns will go for *M. microphyllum*, a mat-forming shrub barely an inch or two tall, with bright green leaves plastered to the contours of the earth. Neither of these has been widely cultivated, but they warrant a place in rock gardens and well-drained xeriscapes.

ASTERACEACE

Patagonia has its fair share of DYCs (damn yellow composites), but unlike the American West, a few pinks and oranges are present in the large-flowered genus of *Mutisia*.

Imagine yellow-orange shaving-brush flowers on a gray-green cushion in a bleak landscape and you will understand the impact of *Brachyclados caespitosus*. The cushions are tight enough in their native habitat to drive even the most staid rock gardener crazy with lust. This plant has not been widely cultivated, if at all.

Chuquiraga aurea is among the most common and largest cushion-forming plants in the central part of the steppe. This spiny aster relative can form cushions the size and height of a Volkswagen bug—impressive! Bright yellowish gold strawflowers lend an air of everlasting flowering to the plant, which is rarely (if ever) cultivated.

As in the American West, gumweeds add life to dry stretches of the steppe. One of the smallest is *Grindelia prunelloides*, with large yellow to orange flowers; flower color in its alpine form is brighter and more saturated, due to the higher ultraviolet light and cooler temperatures. *Gutierrezia baccharoides* is a small mounding to cushion-forming subshrub that inhabits dry parts of the steppe. This gutierrezia has been successfully cultivated in Colorado for several years, probably tracing back to a John Watson collection, and has proven hardy and adaptable at Denver Botanic Gardens. With showy yellow daisy flowers and resinous bright green leaves, it is a noteworthy addition to any rock garden or xeriscape. It bears a strong resemblance to *G. elegans*, first described in 2008 from western Colorado.

Brachyclados caespitosus, one of the brightest flowering plants on the midsummer steppe.

Chuquiraga aurea hails from Chubut south.

The cheery *Mutisia retrorsa* is native to the central steppe.

The richly colored *Mutisia decurrens* is one of the most beautiful species.

Gutierrezia baccharoides near Bosques Petrificados.

Mirroring the flower size and to a certain extent the colors of the South African *Osteospermum*, *Mutisia* is centered in the southern Andes just to the north of Patagonia proper. Several species in shades of orange and pink flirt across the transition zone between forest steppe ecotone and the steppe's western edge. *Mutisia decurrens*, *M. spinosa*, and *M. oligodon* brighten the fences and shrubs they scramble over with orange or pink flowers. All three species have been cultivated to some extent in the northern hemisphere, but none have enjoyed great popularity or longevity. *Mutisia retrorsa* is found only within the central steppe. Like its western relatives, this more arid-adapted species prefers to scrabble through spiny steppe-dwelling shrubs, producing yellow flowers and, later, puffs of white seed. It could prove to be very adaptable to western gardens.

Nassauvia has speciated with the remarkable diversity found in *Helichrysum* in South Africa and in the related *Senecio* in all four steppes. Ranging in form from woody shrubs to tiny alpine rosettes and mats, species occur in almost every environment imaginable on the steppe and alpine regions of Patagonia. Although the genus boasts many potentially ornamental species with interesting forms and beautiful flowers, one stands out: *N. chubutensis* has been cultivated with some luck in private gardens of the Bariloche region. The geometric arrangement of the leaves, together with their long golden spines, creates a truly sculptural plant worthy of inclusion in any garden. The plant is reported to be very rare in the wild.

Nassauvia chubutensis in the garden of Peter Eggert, in Bariloche.

CACTACEAE

A specialty of the New World, Cactaceae evolved after North and South America broke loose from the other continents. The family has two centers of diversity, one located in the high mountainous plateaus of Mexico and the second in the highlands of northern Argentina. Both hotspots are just shy of the two western hemisphere steppes on the side nearest the equator. Some of the hardiest cacti in the world are found in the New World steppes.

A relative of North American *Opuntia*, *Maihuenia patagonica*, the lesser-known and -grown of the Patagonian maihuenias, dwells on the steppe, while the more commonly grown *M. poeppigii* is a Chilean and Argentine montane steppe transition species. Differing in its pinkish flowers (*M. poeppigii* has yellowish flowers) and preference for drier habitats, *M. patagonica* can sometimes be the greenest plant around, especially in the dry *Nassauvia*-dominated steppe near Tecka. In other areas it seems to dominate the steppe, perhaps a sign of overgrazing, similar to *Opuntia polyacantha* on the western Great Plains of North America. Its connection to Colorado is deep indeed as it was introduced to cultivation by local cactus expert Rod Haenni in the 1980s.

Even more *Opuntia*-like than *Maihuenia*, *Maihueniopsis* species from Patagonia may enjoy the notoriety of being the most cultivated Patagonian cacti

Maihuenia patagonica.

Austrocactus patagonicus,
red-spined form,
in Río Negro.

RIGHT *Gymnocalycium
gibbosum* in the garden
of Panayoti Kelaidis.

(apart from *Maihuenia poeppigii*), in Colorado at least. Both *Maihueniopsis darwinii* and *M. ovata* have proven hardy in Denver.

Various *Austrocactus* species are found throughout the steppe (except the far south). Superficially the members of the genus are very similar to the *Sclerocactus* species found mainly in the Great Basin and surrounding areas west of the Continental Divide in North America. *Austrocactus patagonicus* is a very widespread and diverse species, found from southern Mendoza to Santa Cruz. It is occasionally cultivated by specialists in the United States, Canada, and Europe and may have great garden potential in the semi-arid climates of the American West, especially those of the Great Basin and Columbia Basin, which resemble their natural habitat in precipitation regimes.

Pterocactus species, found throughout Patagonia (especially in the drier and warm districts), range from weird and unusually ugly (like a Chinese crested dog) to compact, beautiful, and worthy of inclusion in the garden. *Pterocactus australis* is the most southerly cactus in the world. *Pterocactus fisheri* has probably been the most successful here in Colorado. *Pterocactus tuberosa* has been grown locally, including in the Denver Botanic Gardens, but has not done as well to date. Pterocacti have probably flowered in Colorado only in the garden of Rod Haenni.

Gymnocalycium is widely cultivated in the northern hemisphere. Most species are found farther north in Argentina; one, *G. gibbosum*, makes it to central Patagonia, and it has been cultivated with some success in Colorado. I generally lose it during a hard winter in the garden. With selection, it might be possible to develop a hardier strain.

CALYCERACEAE

Unique to South America, this family holds some of the most bizarre-looking plants in temperate climates. Some resemble flattened cauliflower or broccoli in

Nastanthus patagonicus east of Bariloche, with volcanic ash from Puyehue.

LEFT *Nastanthus patagonicus*, close-up of inflorescence.

leaves or florets, or just a game attempt to mimic a miniature medieval mace. *Nastanthus patagonicus* looks as if it would be perfectly at home on Jupiter but perhaps is better suited to a well-drained rock garden or xeriscape. It has been cultivated in its native Patagonia to some degree.

CUPRESSACEAE

This family spans all four hemispheres, with several important genera reaching into steppe areas and at times becoming the dominant woody plants in those regions. *Juniperus*, *Cupressus*, and *Austrocedrus* are present in three of the four steppe regions. *Juniperus* dominates the steppes of western North America and Eurasia, while *Austrocedrus* borders the western edge of the Patagonian steppe.

Resembling its relative *Calocedrus decurrens* of the Pacific Northwest in foliage and somewhat in form (minus the lovely red bark), *Austrocedrus chilensis* is one of the most adaptable of the Andean trees. It grows in the moist Valdivian forest, where it seems to seek out drier rock slopes, but is also the last Andean tree to peter out on the steppe, where it can form contorted, stunted trees resembling the junipers of the North American and Eurasian steppes. This species, currently cultivated in the Pacific Northwest and Germany, may have more adaptability than previously thought. It has been rated hardy to 0°F and deserves a trial in places like Salt Lake City and even Denver.

Austrocedrus chilensis along the Río Lamy.

PLANT PRIMER

Anarthrophyllum strigulipetalum, close-up of flowers.

RIGHT *Anarthrophyllum strigulipetalum*.

FABACEAE

Along with Asteraceae and Poaceae, this large family appears overwhelmingly diverse at first. It has exploited every terrestrial niche on earth in a myriad of forms and species. The steppe-centered genera *Astragalus* and *Oxytropis* are in this family, and the former at least has a heavy representation in Patagonia. While Patagonian *Astragalus* species are every bit as attractive as their northern hemisphere counterparts, they too tend to be difficult to grow. Several other genera endemic to Patagonia are extremely ornamental and worth any amount of trouble to tame. Other genera are widespread but have very ornamental representatives in Patagonia worthy of inclusion in any steppe garden.

Picking from such a large and potentially ornamental genus as *Adesmia* is difficult, with forms ranging from mats and rock-hard silvery cushions to small shrubs with a spiny, almost geometric growth pattern. Flowers range from peach and violet through to oranges and yellows. *Adesmia boronioides* is singled out because it has been cultivated in both North America and in its native Patagonia. It is also an important medicinal plant for asthma in remote parts of Argentina. It possesses racemes of small but bright yellow flowers above a shrubby plant 3 to 4 feet high with dark green leaves. It seems to be semi-evergreen and is very drought tolerant. The genus as a whole is as ornamental as *Astragalus* and someday, let's hope, will grace gardens around the world.

Little introduction is needed for the famed scarlet gorse of Patagonia, *Anarthrophyllum desideratum*. What is not widely known is that it has at least one additional cushion-forming relative from northern and central Patagonia, *A. strigulipetalum*, with orangey yellow flowers and cushions just as tight and large.

Pink-flowering
Pleurophora patagonica.

Senna arnottiana.

Both species have eluded cultivation to this point. Someday they might be like lewisias, which were once considered difficult and are now commonplace in garden centers.

One of the most exciting things about the Patagonian steppe flora, as with its South African counterpart, is the presence of normally tender-looking things (delospermas, succulent euphorbias) hardy to −20°F. Patagonia promises some of the same with *Prosopis* and *Senna*. Mention mesquite (*Prosopis* spp.), and barbeque comes to mind for most people. Scattered throughout the arid semi-tropical to tropical regions of the world, steppes are not the first place one would look for this genus, yet *P. denudans* reaches Santa Cruz province in southern Patagonia. It is a very ornamental plant, with some forms displaying deep reddish pods and tiny dissected leaves. In the far southern reaches of its range, it is compact. *Senna arnottiana* is an amazing species with large yellow flowers covering a small shrub 3 to 4 feet high; it would look more at home on a sandy beach than the cold windswept steppe.

LYTHRACEAE

The loosestrife family—better known for the medley of variously behaved perennial loosestrifes and crapemyrtle than for steppe-dwelling shrublets—has more

Corynabutilon bicolor, pale form.

Corynabutilon bicolor, dark form.

Corynabutilon bicolor, yellow form.

than these to offer, with at least one long-blooming drought-tolerant shrub: *Pleurophora patagonica*. Few plants are in flower late in the Patagonian summer (most flower in spring and early summer, taking advantage of the higher moisture reserves from winter precipitation), but *P. patagonica* enjoys a long bloom season, brightening the steppe from spring to midsummer with rosy pink flowers over compact shrubs 1 to 2 feet tall and about as wide. It was even in flower in January 2014 in the midst of a terrible drought in Río Negro.

Chloraea alpina.

MALVACEAE

Mallows and their relatives are familiar players in all four steppe biomes, but the genera and their forms vary from region to region. Unlike many northern hemisphere mallows, some in Patagonia are very woody—tiny to mid-sized shrubs with long-lived trunks and stems; and here, where violet shades are rare, these plants are especially welcome, with some of the brightest pinks, violets, and almost blues on the steppe. For example, *Corynabutilon bicolor*, a woody shrub 3 to 6 feet high with ashy gray leaves, has flowers with maroon, purple, greenish yellow, or greenish gray venation. It is nothing short of spectacular and deserves a cult following.

ORCHIDACEAE

Along with Poaceae and Asteraceae, this family shares top prize for diversity within the plant world. More than 30,000 species are known, from deserts to lush tropical rainforest to—surprising to some, perhaps—all four steppe regions. A handful of genera are found in Patagonia proper, and at least two occur not

Chloraea cylindrostachya.

Oxalis 'Ute'.

Oxalis laciniata.

along streams and springs but on well-drained steppe. Of those, *Chloraea* offers spectacular options for the future. *Chloraea alpina* has fragrant yellow flowers with pronouced rippling on the middle lower lip. With *C. cylindrostachya*, tall spikes of green and white flowers rise on a plant 50 to 100 cm high; this superficially resembles *Frasera speciosa* of western North America, and as with so many orchids, the flowers are fascinatingly complex up close.

OXALIDACEAE

This small but widespread family is well known for weedy members that leave a bad taste in the mouth of many a gardener; representatives are found on many continents and within the boundaries of all four steppe regions. Several species of *Oxalis* occur in Patagonian steppe and alpine areas, adding (along with Malvaceae) some of the few violets and saturated pinks to a predominantly white and yellow flora. *Oxalis compacta*, *O. squamata*, and several others have been cultivated with some success in Colorado. *Oxalis adenophylla*, a staple of the Dutch bulb industry, is short-lived in Denver, settling down only for a while. Hybrids that derive from *O. laciniata*, such as 'Ute', have been cultivated more widely in Europe than North America, but none have been particularly long-lived for me.

POACEAE

The Patagonian steppe is full of grass genera shared by all steppes worldwide; however, grasses do not dominate its plant communities to the degree they do in the other steppes. *Festuca*, *Poa*, and *Stipa* (*Hesperostipa*) are well represented, but they and other grasses predominate only in the moister portions of the steppe.

Stipa humilis.

Several *Stipa* species are very showy and have ornamental potential. Of these, *S. humilis* caught our eye as being different. Instead of a few seeds with long plumes (as with most other northern hemisphere needle grasses), shorter plumes are grouped together to form short brushes of furry seeds on a tight compact clump. This grass is especially beautiful when backlit.

POLYGALACEAE

Found throughout the temperate world, this family of approximately 27 genera has an interesting distribution in all four steppe regions. One of the larger genera (700 species by some counts), *Polygala*, does pinks, violets, and blues very well, making up for often small flowers with saturated hues. *Polygala subspinosa*, from the steppe west of the Continental Divide in North America, has striking pink-magenta flowers, and several species in Patagonia have beautiful, intricately structured, blue or pale blue flowers. *Polygala darwiniana*, one of the smaller of these, is a tiny gem for the rock garden or trough. It produces its pale blue to white flowers in the middle of the Patagonian summer, a tiny shot of color in the muted landscape of the central steppe.

Polygala darwiniana.

PROTEACEAE

One knows they are no longer on a northern hemisphere steppe when a certain shrub, with fire-engine-red flowers, is spied. *Embothrium coccineum* occurs

Anemone multifida.

LEFT *Embothrium coccineum.*

everywhere from the moist Chilean side of the Andes to the very dry western edge of the steppe, onto its very fringes, near Bariloche. It is no stranger to cultivation in mild climates, such as the Pacific Northwest and England, but its potential in colder, more continental climates is suggested by its extreme adaptability.

RANUNCULACEAE

Most anemones are woodland or streamside plants, but a few drought-tolerant species dot the steppes of the world. North America has both *Anemone multifida* and *A. caroliniana* in various parts of the steppe. The former is found again in the Patagonian steppe, where it expresses itself with large creamy white flowers (quite different from the typically small pinkish or cream flowers of North American forms); this South American form has been cultivated for some time, often as *A. magellanica.*

ROSACEAE

The Rosaceae is comparatively (relative to South Africa) well developed in the Patagonian steppe. *Geum, Fragaria,* and even *Rubus* are found along its edges, and two genera are centered here: *Margyricarpus* and *Acaena,* which is also present in New Zealand and Australia. Many *Acaena* species that are well established in cultivation are from New Zealand, but several species from Patagonia would make welcome additions to the rock garden. Some form tight silvery cushions; others, textural mats of pinnate leaves followed by bristly seedheads. *Acaena magellanica,* one of the most spectacular, has the ability to form large mats of bluish pinnate leaves and sputnik-like seedheads; these plants are potentially invasive, so caution is advised.

Acaena magellanica
in a ditch near
Gobernador Costa.

Petunia patagonica, DBG.

SOLANACEAE

Outside of the Patagonian Floristic Province, *Benthamiella* species are generally encountered as specialist plants displayed on the show bench in England or tucked into alpine collections in Europe. But this endemic genus may hold some durable rock garden gems or even useful mat-forming plants for xeriscapes in climates more like their native Patagonia. Flowers are in the white to yellow range. Some of the cushions are among the most beautiful and detailed of any Patagonian plant.

Comparatively long established in cultivation, along with *Bolax gummifera*, *Petunia patagonica*, an odd outlier of a genus well known to all gardeners, is unusual in several characteristics: it is perennial, long-lived, woody, and exceptionally cold hardy. It has survived −15°F at Denver Botanic Gardens with practically no damage. With time, the bright green cushion can reach a respectable size and height; large specimens in Santa Cruz were 2 to 3 feet tall and 4 to 5 feet wide. And the amazing flowers, ranging from creamy white to yellow with violet netting, are striking to any lover of unusual plants.

Tropaeolum incisum.

Junellia tonini.

TROPAEOLACEAE

Along with Alstroemeriaceae, Tropaeolaceae is endemic from Mexico to southern Chile and Argentina. The family is composed of two genera and 90 species. Many people know the beloved nasturtium, *Tropaeolum majus*, but not everyone is familiar with the many perennial species found the length of the Andes. Few species have translated well to the harsh continental climate of Denver but at least one, *T. incisum*, makes it to central Chubut and with selection should prove to be a welcome and perhaps vigorous ornamental in Colorado. The flowers are generally a warm and sunny peachy apricot—a wonderful contrast when scrambling over a blue-green mulinum or against the rocky roadsides the plant seems to favor in Patagonia.

VERBENACEAE

Well known in annual and perennial circles, this family reaches one of its evolutionary hotspots in southern South America, where everything from annuals to broom-like trees and woody cushions are found in almost every conceivable habitat niche. *Junellia* is by far the richest genus of verbena relatives in the steppe and a gem of the Patagonian flora. Many species possess an attractive violet or pink color (one is an odd yellowish brown). Only *Junellia micrantha* seems to enjoy cultivation to any degree in North America, but *J. succulentifolia* and *J. tonini* hold promise, the latter species in particular offering a long bloom season of two months. The nomenclature is a bit hazy but will be sorted out shortly (or so one hopes).

THE
SOUTH
AFRICAN
STEPPE

PANAYOTI KELAIDIS

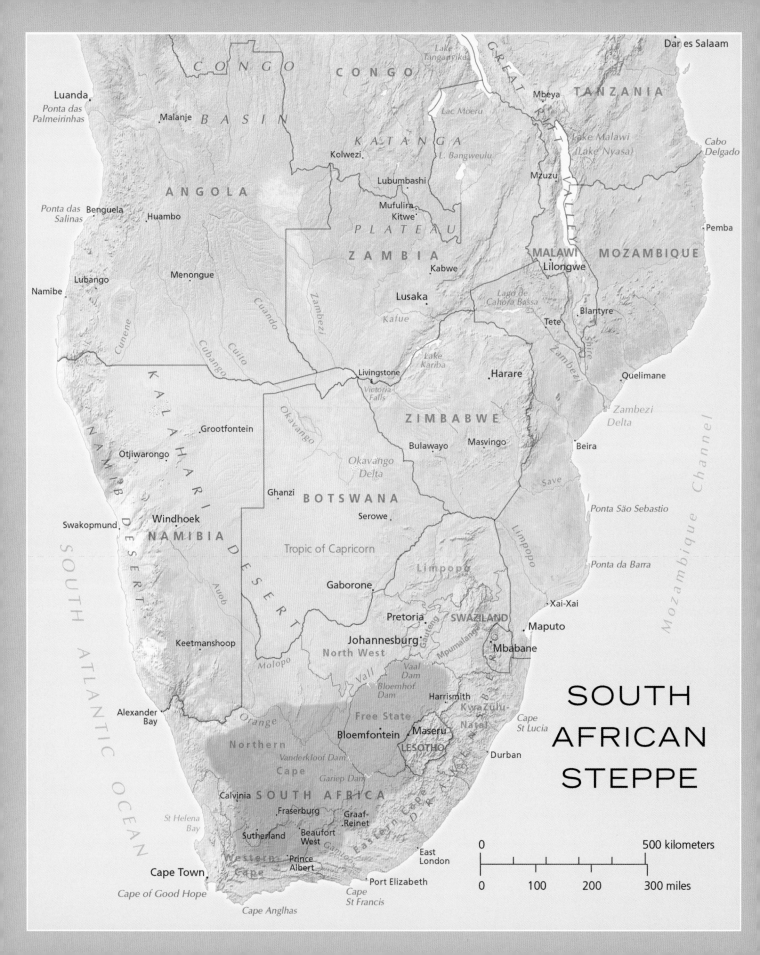

SOUTH AFRICA is half a million square kilometers smaller than Alaska, but it contains more kinds of plants (roughly 25,000 species) than all North America north of Mexico. In fact, South Africa has 40 percent of the plant species found in all Africa, and nearly a tenth of the world's vascular plant biodiversity. The Cape Floristic Province is on the southwestern tip of South Africa, an area roughly the size of California and with almost twice the number of plant species as its sister Mediterranean climate in America. The extraordinary biodiversity of the Fynbos region of South Africa has cast a shadow on the somewhat more diffuse but still very rich biodiversity of the rest of the country. The region in and around the Drakensberg mountains, in the eastern part of the country, is believed to contain nearly 7000 species of vascular plants—making it a botanical hotspot, perhaps not of the same intensity as southwestern Africa, but of greater utility to horticulturists in the northern hemisphere because of the intense winter cold that these highlands experience.

The European settlers who first colonized South Africa came largely from western Europe, so there is very little commerce or communication between South Africa and the other steppe regions. As a consequence, as South Africa grew as a nation, an entire local lexicon was developed by botanists and geographers to describe its landforms and other ecological and biological features. One geographical connection is widely acknowledged: the parallels between the Cape Floristic Province (which is to say the Fynbos, primarily) and the vegetation of its sister Mediterranean climates—in central Chile, southwestern Australia, California, and the Mediterranean basin itself—are blatant. But what is rarely noted in South Africa is the relationship of the more continental climate (encompassed mostly by the Karoo and Highveld) with sister climates beyond the Mediterranean biome.

At this point, it is imperative to clear up an ambiguity that confuses many first-time visitors to South Africa. The vegetation of the Cape Floristic Province is often termed Fynbos (an Afrikaans word meaning "fine woods," or more accurately "fine-leaf woodlands"), an allusion to the characteristic finely textured vegetation that grows on the nutrient-poor quartzite- derived soils of nearly all its mountains and hillsides. In other Mediterranean climate regions, the closest correlative term for Fynbos would be "chaparral," except the soil matrix of most northern hemisphere Mediterranean climates is usually derived from limestone or igneous rocks, which form a much more nutrient-rich soil (in Europe the soil is primarily *terra rossa*, the familiar red clay utilized in ceramic roof tiles and pottery).

Throughout the Cape Floristic Province are wide swaths of land, mostly the valleys between the hard, igneous mountains, where the bedrock was shale. The relative ease with which shale disintegrates compared to the quartzite of neighboring mountains explains the dichotomy of the two predominant vegetation types that occur in this province. Whereas the igneous extrusions weather to form deep, nutrient-poor sandy soils, the shale decomposes into a fertile loam that supported the extremely rich flora characterized as Renosterveld. The past tense in the last sentence was deliberate: as Renosterveld is level and extremely amenable to cultivation, it has been virtually replaced by towns and agriculture. A large proportion of the rarest and most threatened plants in South Africa are the endemics of this biome. Renosterveld has been compared to matorral in the Mediterranean region, but the analogy is fallacious because matorral is not restricted to shale substrates, nor does it represent an assemblage of plants utterly distinct from chaparral. This endangered vegetation type in the midst of the Western Cape is significant because it has so much in common with the more continental Karoo and ultimately the highlands in the eastern part of the country (notably the Drakensberg).

The Fynbos occupies mostly hilly terrain, although there are extensive level areas of acid, quartzitic sand. It encompasses virtually all the classic endemic families of the Cape Floristic Province (Ericaceae, Proteaceae, Restionaceae) and several of the smaller ones (Bruniaceae, Cunoniaceae, Grubbiaceae, and many more). Iridaceae, Thymelaeaceae, and Rutaceae make up a significant proportion of Fynbos flora; these occur in the northern hemisphere as well, but not in such numbers and variety (and with so many distinctive genera). A South African botanist friend of mine has compared the Fynbos flora to a dusty Victorian drawing room with cluttered cabinets full of botanical knickknacks—busy floral ornamentation (think ostrich feathers) piled upon a ground flora that rivals an oriental carpet in complexity. This distinctive assemblage of families and plants is believed to represent relics of an ancient flora that once grew on Gondwanaland—that is to say, its origins predate our current Cenozoic era, tracing back to the age of dinosaurs. Many similar elements persist in Australia and Chile (or as fossils in Antarctica), which strongly supports the notion of Fynbos as a sort of living fossil flora. Since Fynbos soils are far less suitable for cultivating most agronomic crops, especially on the rugged terrain of the mountains, much of this flora is still relatively intact and has been intensively studied by botanists.

The flora of the Renosterveld is closely related to the more inland, subtropical Karoo flora that lies in the rain shadow of the Fynbos mountains: it consists of geophytes, succulents, and all manner of dwarf herbaceous and shrubby plants. The Fynbos's absurd profusion of heathers, proteas, and restios are missing on these richer soils. The families that predominate in Renosterveld are more homologous to those found in steppe climates: Asteraceae, Fabaceae, Iridaceae, Euphorbiaceae, Aizoaceae, and particularly Poaceae, to name a few. Although elements of all these families also occur in Fynbos, they occur much more widely throughout Africa and even beyond, suggesting that these are not unique, Gondwanaland-derived elements but have likely evolved primarily during the current (Cenozoic) era, subsequent to the K-T Boundary event 63 million years ago. The dichotomy between the Gondwanaland elements of the Fynbos and the rest of South Africa is hardly airtight: elements of Fynbos crop up on many of the higher parts of the Karoo. Some botanists argue that, since many heathers, several proteas, and even a few restios are found on the high Drakensberg, these very different mountains might be considered to have Fynbos floristic influence. But these are exceptions that prove the rule.

The flora of the Cape Floristic Province (again, Fynbos, primarily) is so dramatic and unique that it obscures the fact (even for most visiting botanists) that nearly half of South Africa's flora grows elsewhere—and that much of it is indeed concentrated in the Karoo (the term indigenous peoples used for "dry region"), which encompasses Mediterranean and true desert biomes as well as more continental steppe. Far greater in size than the Fynbos, the Karoo is largely semiarid shrub or grassland that from a distance can appear deceptively uniform. For first-time visitors from western America, the landscape is strikingly reminiscent of home: rolling hills, trees confined to valley bottoms—sometimes tree-form willows that look just like the willows in the West. Exotic willows are a serious invasive pest in South Africa, but a half-dozen species of native willows grow there as well.

The Karoo is not only vast in area—occupying nearly two-thirds of the country—but extremely diverse in habitat, microclimate, and landform. Much of it is subtropical: the Little Karoo, the Robertson, and Tankwa Karoo experience only superficial frost most years, which does not severely damage frost-tender plants (hence the vast citrus orchards near Worcester, for instance). These warmer

Salix mucronata near
Somerset East.

Typical Karoo vista near Uniondale, Eastern Cape.

valleys, the most accessible parts of the Karoo bordering the Cape provinces, are also the most botanically rich. They are the close correlatives of the central valley of Chile or California—facultatively and practically extensions of the Mediterranean climate region that dominates the more coastal Fynbos, or better said, the coastal Renosterveld.

Many botanists also characterize the vegetation in the warmer Karoo as Renosterveld—and the parallels with the remaining fragments of Renosterveld in more coastal regions are striking and support this usage. At one time the uncultivated, pristine Renosterveld would have suggested little islands of Karoo interdigitating, as it were, the distinctive Fynbos—something that wall-to-wall farming and the spread of towns and cities in the Cape provinces has truly obscured. Perhaps because so few extensive areas of Renosterveld persist in the more coastal mountainous areas, the true relationships of more coastal Renosterveld and its much more extensive Karoo flora are not as evident, nor are their distinctions as easily limned.

Geology and Climate

The geology of southern Africa (South Africa, Lesotho, and Swaziland) is undeniably complex, with nearly 700 distinct ecological vegetational units based on the geology, soil, and climate (Mucina and Rutherford 2006). Nevertheless, South Africa has had far less geological activity for millions of years compared to the dynamic mountain-building activities in western America and Eurasia. There are a few regions where limestone outcrops in South Africa, and sandstone occurs in distinctive outcrops throughout the Drakensberg and southwestern Africa as well. But the enormous batholith that underlies South Africa, emerging as igneous rock in much of the country, and as the black basalt of the high Drakensberg, represents one of the most geologically stable and ancient, undisturbed portions of the earth's crust.

The sterile quartzitic soils that characterize Fynbos and the richer shale deposits that prevail in Renosterveld and parts of the Karoo have been touched upon. But the soils throughout much of the Karoo and Drakensberg derive from nutrient-rich igneous and volcanic stone that decomposes and forms a deep, rich loam not unlike the soil we yearn to have in our vegetable gardens. It is perhaps not an accident that so many South African plants thrive in cultivation, when they evolved in a soil matrix so similar to what we attempt to provide in most home gardens.

As is the case with the Patagonian steppe, South Africa's steppe experiences far milder climatic extremes than northern hemisphere steppes. Although winter and severe frost occur at lofty altitudes throughout the central parts of the Karoo and Drakensberg, temperatures rarely drop below 20°F for extended periods as they do every year in much of the American West or Central Asia. Likewise, temperatures in excess of 86°F are not nearly as common in much of the Karoo or Drakensberg steppe, although northern South Africa can get extremely sultry as one approaches the tropics and drops in altitude.

The characteristic cline of all steppe regions holds true in South Africa, with winter rainfall concentrated in the most southwesterly corners of the country and summer rainfall prevailing in most of the inland Karoo and the eastern parts of the country. The winter rainfall usually comes from the Antarctic in gusty storms that are drizzly and last for many days, while the summer rainfall is generally produced by warm, convectional thunderstorms that spiral from the Indian Ocean. In some years the diurnal storm cycle can produce abundant rainfall in heavy rain showers every afternoon, always clearing by evening; in others, the clouds form without much rainfall. The pattern varies drastically from year to year.

GREAT KAROO

The bulk of South Africa's interior is referred to as the Great (or Nama) Karoo (a word derived from the San, for "dry"). Although some parts are so mild that hardier eucalyptus serve as street trees, most areas experience frequent and severe frost; winter temperatures consistently dip below freezing, so that citrus and olives or the more tender Mediterraneans succumb. The higher escarpments that encircle the Great Karoo (the Hantamsberg, Roggeveld, Komsberg, and Nieuwveld mountains) experience even more severe winter cold and dustings of snow; temperatures regularly drop to −10°C in this region, and severe frost can occur from May to November. By any measure, the high plateau that is the Great Karoo qualifies as cold temperate and constitutes the closest climatic parallel to the steppes of South America and the northern hemisphere.

Much of the Great Karoo has been heavily grazed by sheep—some of it excessively so. Some botanists believe that the "succulent Karoo" found in the drier, more subtropical areas (such as the Tankwa) has been spreading eastward in the last century, replacing what had been grassland lost to overgrazing and poor land use. In a good spring, following ample winter rainfall, grass returns to predominate throughout, heavily intermixed with an extraordinary abundance of bulbs, annuals, and perennials of all kinds. In many parts of the Great Karoo, as in the sagebrush steppe of western America, shrubs can dominate the landscape. Most genera are in the Asteraceae; those most frequently encountered (*Euryops* among them) have tiny leaves and small flowers.

This vast region is occupied by enormous ranches and has some of the most extensive mining activity in the country. It is the least studied part of South Africa, usually dismissed as being devoid of botanical interest—in truth, no one has bothered to look much—but there are in fact hundreds of endemic taxa, and likely far more remain to be discovered.

EASTERN CAPE KAROO

The principal escarpment of the Drakensberg tapers dramatically as it crosses the border from Lesotho and KwaZulu-Natal province into the Eastern Cape proper. Gazing southward from Tiffindell (South Africa's most ambitious ski resort, just inside the Eastern Cape) to the west and south, you see a whole archipelago of mountain ranges: these sky islands of the Eastern Cape harbor a flora similar to the high Drakensberg on their summits, but their foothills and the deep valleys between them harbor a prairie-like steppe flora of great diversity, which in turn grades into subtropical near-desert or evergreen subtropical thorn forest as you descend in elevation. The colder, steppe flora extends to high levels on these peaks on exposed slopes. Many plants that can be found in the Eastern Cape steppe (*Pelargonium abrotanifolium*, for example) are the same as those found throughout the grassy portions of the eastern Karoo. This is an area of tremendous biological complexity and diversity: many significant groups of plants, including the succulent *Bergeranthus*, have their centers of diversity in or near this region. The spectacular *Delosperma* Fire Spinner was collected in this area on Sneeuberg, the highest peak in the Cederberg range.

High-altitude grassland near Balloch, Eastern Cape.

THE HIGHVELD

Highveld is the term applied to the somewhat more humid grasslands that prevail from the Great Karoo to the Drakensberg, especially to the north around Johannesburg, including the foothills and meadows higher up in those mountains. Many familiar genera of grasses (*Themeda*, *Merxmuellera*) are found in this region, which is a close climatological correlative to the grasslands of the lower U.S. Midwest.

THE DRAKENSBERG

The Drakensberg would at first seem to be too high and too wet to be classed as steppe, and much of the mountain range is undeniably humid and different from the sparse grasslands found farther west. Prevailing summer monsoons, the source of most precipitation, drop their moisture on the steep, east-facing escarpment of both the northern and southern Drakensberg. Extensive forests of yellowwood (several *Podocarpus* spp.) and a rich tree flora are still abundant in deep ravines here, and even the rugged slopes grow lank with much taller bunchgrasses

Inulanthera calva, close-up.

RIGHT *Inulanthera calva* and *Eumorphia sericea*—two widespread shrubs in the Drakensberg, carpeting the foreground of the bluff at Oxbow, Lesotho.

and shrubs: these high meadows of "sourgrass"—dominated by *Themeda triandra* and *Merxmuellera* species—are strongly reminiscent of the tallgrass prairie of the U.S. Midwest. The landscape quickly reverts to woodland if the tall herbs are burned or overgrazed. To the west the landscape is considerably drier, with some valleys (the Quthing Gorge in Lesotho, for example) falling in such rain shadows that they approach truly arid conditions.

Grasslands constitute the bulk of the Drakensberg; in this, it is similar to many mountains in western Asia, which have continuous grassland steppe vegetation from their lowest foothills to their alpine peaks. The grasslands here have a great deal in common with the Karoo, in terms of species composition and appearance. Many of the predominant plants (*Ruschia hamata*, *Euryops* spp.) are the same in the true Karoo at the base as on the slopes extending to the very summits of the escarpment. Large areas can be dominated by such shrubs as the yellow-flowering *Inulanthera calva* and white-flowering *Eumorphia sericea*.

A very eloquent illustration of the relationship of Karoo and Drakensberg flora is *Delosperma cooperi*, probably the most widely cultivated South African succulent in the northern hemisphere. This species is abundant on the steppe over large areas of the grassy plains of the Free State. I remember seeing yard-wide mats of it in full purple bloom, dotting the landscape in both directions, as I drove from Bloemfontein to Lesotho; at the time, I assumed this taxon was restricted to these lower elevations and interior grasslands—that is, until I spent several days at Oxbow, Lesotho's principal ski resort (a rough one, it's true: no chair lifts, for one thing). Nearby, the steep, north-facing slopes (which are the sunnier ones in the southern hemisphere) of an 8000-foot pass were draped with huge mats of *D. cooperi*, very similar morphologically to the form growing 4000 feet below on the plains. I have also seen *D. cooperi* growing on Blue Mountain pass, and it is recorded extensively in the foothills around Roma, so its range could be described as montane steppe in the Drakensberg and grassland steppe in the Karoo proper.

Plant-People Connection

South Africa has been continuously populated by humans from our very origins as a species: the San (or !Kung) and Khoi peoples are believed to be descended from the original inhabitants of the region. These slender, delicately built humans with light bronze skin are as distantly related to the much darker Bantu as they are to Europeans (the two groups who now constitute the bulk of South Africa's population); mitochondrial DNA has suggested they may in fact be ancestral to all humans.

Since the history of human occupation goes back millions of years in South Africa, it is impossible to imagine what the flora might have been—or might be—without human impacts. For instance, when the very first hominids evolved to live primarily on grassland on or near South Africa, lions, leopards, cheetahs, and several other large, now-extinct carnivores shared their space. Anthropologists postulate that we humans evolved our larger brain, our intelligence, and especially our social organization in response to the fearsome threats our delicate ancestors faced in this harsh environment. The elimination of these carnivores (the familiar three are now mostly restricted to the Lowveld) obviously had enormous impact on the herbivores they preyed upon, and consequently the landscape grazed by the herbivores.

It has been postulated that the extinction of a half-dozen predators might have been a deliberate consequence of competition with humans for food—and the humans may also have sought to eliminate their competition by hunting them as well. Conversely, the fact that far more large herbivores have survived into the Holocene in Africa than on the other continents is powerful evidence that the rich megafauna of Africa has persisted due to coevolution with humans. As *Homo sapiens* emerged from Africa onto Eurasia, and ultimately the Americas and Australia, we eliminated a far greater number of the largest animals, which had not evolved to evade a bipedal predator.

Is South Africa therefore perhaps closer to a "pristine" state than the other continents as a consequence of this coevolution? Most researchers believe the reason grassland is so prevalent in Africa (less than 5 percent of South Africa is naturally wooded) is likely down to the enormous grazing pressures of once-vast herds of ungulates, in combination with frequent grassfires set by thunderstorms (and likely by primitive humans as well)—the same pair of circumstances that drove the tallgrass prairie of the U.S. Midwest into being. Throughout both regions, indigenous peoples used fires to provide fresh pasture for their animals.

As everywhere on earth, the impact of humans since European and Bantu influx into South Africa four centuries ago has been enormous and continuous. Much of the land suitable for farming has been plowed, and livestock grazing is universal throughout the country. I suspect that the "natural" appearance of much of South Africa today could well be a result of a landscape that has developed enormous resilience to humans over millions of years, and to the constant hammering of herbivores—wild or domesticated.

Touring the South African Steppe

SOUTHERN DRAKENSBERG

Grass is the unifying note across the entire landscape of eastern Africa. The grass turns yellow, tan, and auburn by March most years and may not green up at higher altitudes until November—almost nine months of dormancy—indicating that this is indeed a temperate climate. But from October to February—given average rainfall—the grasslands are intermingled with an astonishing variety of perennials, bulbs, and shrubs, which creates an almost gardenesque picture, especially since so many of these have been adapted to north temperate perennial borders. In early spring the white flowers of *Anemone fanninii* are common in the foothills (an interesting link to a predominantly north temperate genus); this species is a giant among anemones, with huge leaves and flowers 5 inches or more across.

There are slopes throughout this region where representatives of *Agapanthus, Crocosmia, Gladiolus, Diascia, Geranium*, and more intermingle, much as they would in a lush English border. Philosophical visitors can hardly be blamed for speculating that there must indeed be a very deep operational memory in mankind: have we fashioned gardens on this wild model, where we might have originated as a species?

As you cross the Karoo from west to east toward the Drakensberg, the landscape becomes lusher and greener with more grasses. It is reminiscent of driving east through western Wyoming, past Casper, until one reaches the full-blown grasslands of the Great Plains. The Free State is the southern extension of the Highveld centered around Johannesburg, a few hundred miles to the north. It is a landscape very much like the Front Range of the Colorado Rockies. Instead of prickly pears, however, you will find several kinds of ice plants. *Delosperma floribundum* forms pink pools of color much of the summer along roadsides near Springfontein. Straggly mounds of shrubby *Ruschia hamata* can be a foot or more tall and wide. *Chasmatophyllum musculinum* has a wide range along the base of the Drakensberg foothills, a tiny, jagged-leaved clump-former with bristly,

Brunsvigia radulosa on Eastern Cape grassland.

night-blooming flowers of golden yellow; in cultivation this can start blooming in April in the northern hemisphere, lasting until November in a milder year—surely one of the longest blooming perennials!

As in Colorado, any cluster of rocks on the veld seemed to have ferns poking forth—usually *Cheilanthes eckloniana* (in the Great Plains this would be the similar *C. feei*). Rather than yuccas, aloes occur here and there at the lower altitudes: *Aloe broomii* dots the prairies for miles around Aliwal North, for instance. Although the giant brunsvigias with waxy, scarlet flowers much like a hippeastrum generally bloom in March, in the southern Drakensberg they can bloom over a much longer season (I have often seen plants blooming in February and even January: the flowers were on foot-long pedicels, which formed a shimmering, perfect sphere). Both *Brunsvigia radulosa* and *Boophone disticha* are tumbleweeds, scattering their already germinating seed across the veld. *Brunsvigia natalensis*, growing above 8000 feet on the rim of the Drakensberg in KwaZulu-Natal, occurs in both soft pink and carmine-red. High-altitude germplasm would likely be hardy in northern hemisphere cold-climate gardens.

The stark Drakensberg landscape in Lesotho (an independent kingdom roughly half the size of Colorado) shows the wear and tear of overgrazing: its people, the Basotho, were nomads until the 19th century and have always been herders. Each village seems to have a throng of children who are attracted like magnets to the few cars that drive through. The most notable plants (aside from a few extensive plantings of corn and other vegetables) are gigantic stands of *Agave americana* that ring every village, making vast forests of flowering stems in their midsummer. The Maluti Range parallels the western boundary of this land-locked country—the first high range of the Drakensberg tending from southwest to northeast, not far from Maseru, Lesotho's capital; and a perfectly serviceable road climbs up the somewhat overstated "God Help Me Pass," topping out at roughly 7000 feet. The vegetation over the pass appears to have been heavily overgrazed, although on the summit, some lovely montane plants persist. There is not a lot of woody vegetation in evidence, aside from tufts of *Leucosidea sericea*, a silvery-leaved rose family plant that forms thickets in many gullies. The first Drakensberg alpines one might find here include sizable clumps of the ubiquitous *Kniphofia caulescens*, giant *Dierama robustum*, easily 7 feet tall, with lovely pale purple bells, and *Zantedeschia albomaculata*, growing in moist swales: all classic Drakensberg wildflowers.

Crossing the pass, one drops back down to a steamy and surprisingly subtropical valley. The vegetation grows sparse again (vegetation is cropped short everywhere cattle or sheep can be herded conveniently within a half-day's walk from their kraal). Little children once again throng around any passing car. As the road climbs up on to the lusher Blue Mountain, a very different scene unfolds. Steep green peaks tower all around; waterfalls abound on all sides, falling for hundreds and hundreds of feet. Dead carcasses of buses and a few cars have been left perhaps deliberately to caution motorists. And suddenly a whole mountainside begins to emerge behind the shoulder of a green hill—a mad, expressionistic canvas daubed with bright swaths of primary colors. Much of this range had been burned the year before, and the ground beneath the charred *Leucosidea sericea* and *Buddleja loricata* was a dense tangle of maybe a hundred species of incredibly showy wildflowers forming a dense mesh of bloom.

The spectacle that greets a lucky visitor on burned slopes like this exceeds even the lavish displays one can reliably find on Sani Pass or Mont-aux-Sources. The brightest hue was undoubtedly *Diascia anastrepta*, a deep salmon-pink twinspur in the snapdragon family that forms dense mats a meter or more across with flower stems up to 15 inches tall—a solid carpet of color. Yellow is provided by bright yellow gentians (*Sebaea* spp.) and large mounds of a giant-flowered berkheya, one of those spiny asteraceous plants that looks like a cross between an artichoke and a sunflower. Violet comes courtesy of innumerable plants of the wonderful lilac pinky purple *Senecio speciosus*, found throughout the Drakensberg. Scarlet flashes emanate from several species of *Zaluzianskya*. Wherever one looks, there are orchids, ferns, and stonecrops; perhaps six or seven *Crassula* species occur on this slope alone—tiny groundcovering *C. setulosa* and husky biennial monsters (*C. vaginata*). Five or so genera of composites with pure white flowers include *Helichrysum*, which alone might have a dozen species represented here. And everywhere the chartreuse pompons of *Scabiosa drakensbergensis* are in evidence.

Only a few minutes hike from the burned slopes below, an entirely different assortment of plants stretched in all directions from the top of the pass. Climbing toward the north and west, three or four species of heather tangled the moist peaty hollows. A mounding *Selago flanaganii* with cool lavender spires—like some sort of glorified, glowing veronica—spread into mats a foot or more across and up to 6 inches tall. Dense mats of *Hirpicium armerioides*, an alpine ecotype much tighter than what we have grown in the past, resemble those of a dark blue-green armeria.

Three or four white helichrysums appear (the showy, large-flowered *Helichrysum bellum* and *H. marginatum* growing everywhere here)—then the first giant-flowered yellow everlasting—the wooly-leaved *H. aureum* var. *scopulosum*. Only a few were in fresh bloom, but literally thousands all the way up the mountain were in full, puffy seed. Next *H. sessilioides* appears on a cliff face. Every crevice on one particularly large igneous boulder is full of giant wads of this stunning cushion alpine, in full bloom. More and more crassulas—in bloom, in seed. Mystery scrophs, more orchids, and suddenly the first alpine ice plant shows up—a tiny cool pink delosperma—species unknown. A giant pink-flowered delosperma, like a clump-forming *Delosperma cooperi*, grows right next to a pulvinate *Ruschia putterillii*. Not far away, a shrubby crassula appears—a yard tall, two yards across. This one is in full bloom—a glistening ivory-white. This is the shrubby *Crassula sarcocaulis*, no doubt, a frost-adapted jade plant. There's a flash of coral some 20 feet away. One spies the giant, waxy leaves of a succulent with stems a foot or so tall with waxy, orange-red bells: *Cotyledon orbiculata* is one of those universal succulents that can be found practically anywhere in South Africa from the Atlantic coast through the mountain Fynbos among proteas, across the rocky deserts of the Karoo. And here it was on the very top of the Drakensberg at almost 9000 feet.

Words cannot convey the magical light on the mountaintop, the vista of

Biennial *Crassula vaginata* on the Sentinel, Free State.

Cotyledon orbiculata,
close-up of flower.

central Lesotho fading into the distance, from jade, lime, and a ludicrous Irish green into the blue and bluer distant peaks of the KwaZulu-Natal escarpment. Every few feet, some new mystery appears. Dozens and dozens of stunningly beautiful ornamentals grow more thickly even than in a garden.

PLATBERG

The four-hour drive from Maseru to the charming town of Harrismith runs along the Maluti Range for about the same distance that Interstate 25 parallels the Front Range from Denver to Raton; but while literally hundreds of botanists have combed that area in Colorado (albeit not nearly as thoroughly as European ranges), gigantic stretches in the Maluti had never been trod by scientists until recently, when roads to service gigantic hydroelectric projects were built throughout the range. The narrow canyons of the Maluti are the principal habitat for the endangered *Aloe polyphylla* (spiral aloe), the national flower of Lesotho. Researchers have watched the numbers of this aloe plummet—mostly thanks to vandalism by bored shepherds. The rolling plains along the base of the Drakensberg—and even the sharp escarpment of the mountains in the distance—possess the quality of light one finds in the Rockies or driving along the Tien Shan in Kazakhstan. These could be grassland in southern Idaho, say, and the magnificent mountains to the east look very much like the Sawtooth Mountains, driving along the montane steppe north of Boise. The second you get out of the car, however, the illusion is shattered.

Platberg, near Harrismith in eastern Free State, is a floral paradise nearly 20 miles across—fenced and protected for the sake of the wildlife. It is just one of thousands of "table mountains" (*platberg* in Afrikaans) that occur around all the Drakensberg and north to the Highveld—much as mesas and plateaus characterize the American West. Steep cliffs—hundreds of feet—protect the top, except at a few access points. The climate must be very nearly the same as Lubbock, Texas, say, or Tulsa, Oklahoma—only not nearly as hot or humid in the summer.

It is January, and the profusion of flowers here is equal to that on Blue Mountain—maybe even richer in variety, if not display. The first new taxon is a giant-flowered, prostrate ipomoea with wine-red flowers 6 inches across. This bears an uncanny resemblance to *Ipomoea leptophylla* of the Great Plains—the first of many uncanny parallels here. Alongside the ipomoea four or five different kinds of bulbous plants have already gone to seed—probably from *Tritonia* and *Gladiolus*, perhaps *Cyrtanthus*. One notices a luxurious, lax mat-forming plant, looking like a cross between *Ruellia humilis* and an ajuga. This is *Barleria monticola*, probably the likeliest South African member of the Acanthaceae to be hardy; it would be a wonderful border plant, with dark blue-gray leaves and giant blue petunia flowers.

Montane mesa near Tarkastad—typical of plateaus throughout eastern South Africa.

Composites everywhere—the one that first strikes me is a relatively minuscule vernonia only a foot or so tall with downy leaves that remind me of an anaphalis, only with deep violet-blue clusters of flowers on top. So many South African plants bloom in the summer—prompted by heat and rainfall—that their bloom periods often extend for weeks at a time.

Gradually trees close around, and one can only glimpse pieces of Harrismith here and there through the yellowoods—much as one glimpses Boulder on a hike up to Royal Arch on Green Mountain. Do not be surprised if blood-curdling screams echo wildly at this narrow point of One Man's Pass: it's just a troupe of baboons in the middle of a family squabble. When your heart starts beating again, you may notice sheafs of *Silene undulata* with wonderful white pouchy

One of many *Vernonia* species found throughout South Africa—a genus that occurs in east Asia as well as North and South America.

flowers growing mostly in shade. And then, still in the dark of the trees, the first *Streptocarpus gardenii*, cool blue-violet long-tubed trumpets with a lax rosette of giant leaves arising from the scree. How amazing to see plants I always thought were tropical in such a temperate setting!

A little higher and the first miniature kniphofia shows up: a bright orange form of *Kniphofia porphyrantha*. Little more than 15 inches tall, with a bristling huge flower, it positively glows in the half-light of the chasm. Higher still, and there on the rocks is a bright magenta flash—ice plants! This deep pink form, similar to what is grown as *Delosperma ashtonii*, is tucked perfectly into crevices here and there. Hesperanthas bloom alongside it in the crevices. The sky begins to open up, there are daisies of all sorts—you are on top of the Platberg.

The summit is not flat at all but a series of billowing hills, looking a little like the ocean from a dinghy. And every foot or so there is something new. Overhead, puffy white clouds float past—a perfect Nebraska sky. A cool breeze makes walking effortless. Helichrysums everywhere—who could begin to key them all out! *Helichrysum adenocarpum* is easy only because it is among the few that come in pink. One of the commonest on alpine cliffs and screes is *H. milfordiae*, which

Helichrysum adenocarpum on Platberg.

LEFT *Helichrysum milfordiae* is common on alpine cliffs throughout the Drakensberg, and well ensconced in cultivation.

has been cultivated in Europe and America since the 1940s; The tiny and delightful *Gazania linearis* throngs on steep gravelly slopes, some plants with a dozen flowers open simultaneously. In places this makes a perfect Persian carpet design mixed with deep blue and purple nemesias on stems only 8 inches tall.

A pond lies just above the crest of a hill, ringed with a series of small outcrops. Sheets of daisies are here—*Helichrysum chionosphaerum*, in its clumping rather than the commoner mat-forming phase, delights with its perfect bubble-like flowers. Shrubby *H. trilineatum*, a yard tall. Strange succulents show up on the drier rocks; densely tufted ruschias just finished blooming—8 to 10 inches tall and 2 feet or more across: seemingly intermediate between *Ruschia putterillii* and *R. hamata*. Numerous species of *Crassula*, some of them solitary clumps, others forming tufts, and some making shimmering white mats. Incredible variety of bulbs—*Gladiolus*, *Moraea*, a deep yellow bulbine with flowers the size of a hyacinth: perhaps *Bulbine abyssinica*?—who knows what they all are.

In full, pure yellow bloom, *Aloe ecklonis* in a miniature phase dots the top of one sparse hillside; this deciduous aloe has overwintered in Colorado on several occasions but not yet reliably. *Aloe aristata* with its lovely red tubular flowers was only just passing over; this makes cushions a foot or two across. Strange to see an aloe growing so high—densely tangled with ferns and grasses, like a sempervivum in the Alps. *Euphorbia pulvinata* grows here as well, each head almost 2 inches across—hundreds clustered on many plants; this makes a much flatter specimen than *E. clavarioides*.

A day on the Platberg flies by, cameras practically smoking from all the pictures! Time to head home: on the descent of the west face, one passes an entirely different assortment of plants. *Dianthus basuticus* in a particularly giant-faced ecotype, a bright pink carried on stems less than a foot tall, forms huge mats on the cliff—looking much like a wild cottage pink in Europe. *Hirpicium armerioides* here forms lax mounds of foliage three or four times wider and taller than the alpine ecotype. A giant white monsonia flowers on stems 6 inches tall or more; it was lovely in drifts here and there. The true spectacle of the mountain, however, appears only now: the entire west face of the Platberg is gauzed with bright blue from a distance: masses of *Agapanthus campanulatus* ssp. *patens* in full, midsummer bloom! Big clusters of bright blue lilies, on stems a foot or two tall, over clumps of strap-shaped leaves. They formed drifts on the cliffs, grew out of the rock face, in the grass, on the scree. This must be one of the finest sights one can see on this very flowery mountain.

MONT-AUX-SOURCES

Mont-aux-Sources—the second highest peak in the Drakensberg—forms the boundary between two of South Africa's nine provinces (Free State and KwaZulu-Natal) and the Kingdom of Lesotho. The approach to the mountain with the romantic French name is through the former bantustan of QwaQwa. As is the case in so many of the so-called homelands, the land is overgrazed, the villages very poor, and the amorphous city of Phuthaditjhaba, many times the size of Harrismith, sprawls over much of the now surprisingly urban "homeland." Base camp for Mont-aux-Sources is the parking lot at the Sentinel—a stark giant monolith that appears to have drifted a quarter-mile or so north of the gaunt Drakensberg escarpment.

Any wildflower lover climbing this mountain in January will be transfixed by the stunning variety of plants, even on the way up. For the last few miles of the drive from the Witsieshoek Mountain Lodge to the Sentinel, steep cliffs along the east side of the road are draped with immense cushions of *Helichrysum sutherlandii* just coming into bloom. Several dozen species occur on this mountain alone, only one of which (*H. retortoides*) was in seed on our midsummer visit. A miniature *Crocosmia pearsei* with giant orange and scarlet flowers on stems a foot or so tall grows right next to the parking lot. Many species of waxy white, yellow, or gray asclepiads form tiny mats and tuffets; there are deep violet or pale pink geraniums (including *Geranium drakensbergense*) and endless, infinitely variable composites. The road terminus is already above 7000 feet; from there, a wide, beautifully maintained trail rises for several miles to the summit plateau. The last few hundred feet are so steep that two chain ladders side by side are used to scale the sheer cliffs—an experience that is guaranteed to thrill anyone who's not a seasoned mountaineer.

An astonishing range of herbaceous plants crowd both sides of the trail, all the way to the summit; bulbs are especially profuse along it—six or seven *Moraea* species alone. *Moraea modesta* grew from cliffs, with full-petaled roundish flowers almost 2 inches across speckled with white and deep violet. *Moraea trifida* grew nearby in virtually the same fashion, only with spidery, pale whitish flowers. I was excited to find my first dense colonies of *M. alticola* with their huge clumps of glossy leaves; these had largely finished blooming, although I did find a few fresh, giant straw-yellow flowers. The loveliest moraea on the mountain must have been *M. inclinata*, with sapphire-blue flowers on wiry stems 8 inches tall or so; this grew by the hundreds in meadows as I returned—I hadn't seen a single one in the morning. For the miniaturist, *M. alpina* grew only an inch or two tall—wonderful pure blue flowers. On subsequent trips I found *M. robusta* with giant yellow flowers on stems just a few inches tall, and a mysterious species that was determined a year later to be new to science: *M. vigilans*.

All along the footpath, more and more varieties of everlastings could be found—dense mat-formers with pinkish flowers, yellow-flowered buns, deep green twiggy groundcovers. Once on top, prepare to be astonished by an entirely different assortment of alpines. Many kinds of orchids (including *Brownleea*) cling to cliffs, and dozens of *Disa* species punctuate the "grass" (*Hypoxis* and an almost solid carpet of *Rhodohypoxis*) in all the low seepages on the top, making pink and white speckled tapestries on the ground. The display of monocots is reminiscent of the Karoo on a good spring day.

The monocots that painted the brightest picture on the mountain, however, were the graceful irids of *Dierama*. These usually form dense tufts with giant wands of bloom 6 feet tall or more in pale pink, white, or lavender. A few giant clumps occurred on the higher slopes of Mont-aux-Sources, but the commonest dierama on this mountain is a miniature species usually just a foot tall—sometimes growing up to 18 inches. This is *D. dracomontanum*, a recently described species that comes in a dusky red tone, the color of tomato soup or Bryce Canyon at dawn. Rather than growing in solitary clumps, as the larger sorts do, these occur in tremendous quantities, thousands of plants growing cheek by jowl, so that from a distance the flanks of the Sentinel are visibly tinted pink. This same dusky red—otherwise rare in the plant kingdom—is also prevalent in *Diascia*.

The composite that makes the greatest impact is the deep rose-magenta cousin of *Osteospermum jucundum*, the alpine "Freeway Daisy" that occurs throughout the Drakensberg—always above 7000 feet. On the Sentinel cliffs, a local race forms mats 2 feet or more across, deep green oval leaves studded with giant flowers almost 3 inches across; most plants appear to be identical to what British horticulturist Helen Milford described in the 1940s, in the Alpine Garden Society's bulletin, as "*Osteospermum barberiae compactum*." The true *O. barberiae* is a lowland plant that is intensely aromatic—and lovely in its own right.

Brownleea macroceras, one of 300 kinds of orchids that occur in the Drakensberg alone.

Drakensberg osteospermums are very variable; there were wonderful pure pink forms farther south, and a glistening white phase on the Platberg across the valley, but the glowing magenta-purple race from the Sentinel will always be the ultimate representative of this remarkable ornamental.

Proteas and More

The family Proteaceae is largely confined to the southern hemisphere—mostly South Africa with a large secondary center of diversity in Australia and then South America, with one or two renegades veering northward beyond the equator. *Protea roupelliae* has perhaps the largest flowers and most stunning coloration of the Drakensberg species; it seemed to be just a shade smaller than the more famous *P. cynaroides* (king protea) of the Cape provinces, the national flower of South Africa. Trees of it were everywhere, to almost 8000 feet in elevation! In more protected gullies *P. subvestita* formed a more slender tree with silvery leaves and smaller flowers that stay vase-like, never opening into such a wide chalice. The last arboreal protea, *P. caffra*, usually grows at lower elevations, with dark leaves and deeply colored flowers. The presence of these seemingly subtropical plants at high altitudes in the Drakensberg highlights the complexity of these mountains—and the futility of generalization. It is unlikely any of these larger species will succeed in cold-climate regions (I suspect few have been attempted), but those that grow at higher elevations, where snow and frost is sufficiently severe, might be adaptable to parts of Arizona or New Mexico.

There are, however, several species of short, shrubby proteas at high elevations throughout much of the Natal Drakensberg. These appear to be absent from the drier, western mountain slopes, but a mile or two south of Oliviershoek Pass, the sparse grassland was interspersed with mounds of *Protea dracomontana*, the commonest of the low species. Some plants were more than a foot tall here, like somewhat scruffy manzanitas, growing in soil that resembled gumbo. The flowers, however, were almost 4 inches across; plants had already been blooming for months and looked to have many buds coming along. The intricate, jewel-like blossoms varied from purplish garnet to pale tangerine and peach. *Protea compacta* generally grows at lower elevations and is somewhat similar in habit. Here the flowers tended to more pearly shades of pink and white. Both species burn regularly to the ground during the dry winter months and regenerate from an underground caudex, which bodes very well for their winter tolerance in our somewhat more variable and extreme climate. So even this family, so characteristic of a maritime climate, has slowly adapted to harsh, subalpine environments in the South African steppe, far from oceans in the northern hemisphere.

A few decades ago, when Denver Botanic Gardens was still in its infancy,

Protea roupelliae
blooming in the high
Drakensberg in January.

the only South African plant routinely seen in gardens was a kniphofia, usually labeled *Kniphofia uvaria* (although it is likely a hybrid involving several other species). Today, dozens of species of South African plants are used in regional gardens, and possibly hundreds are on display. Denver Botanic Gardens was largely responsible for this horticultural revolution, insofar as garden staff brought so many of these plants into cultivation, and the gardens publicized these plants by displaying them prominently. Looking back, I wonder how much our curiosity about steppe wasn't inspired by the diverse steppe of South Africa.

SOUTH AFRICAN STEPPE

AIZOACEAE

The distributional ranges of several genera of ice plants are centered in the eastern portions of the Karoo, extending onto the foothills of the Drakensberg, and in a few cases to the cold tundra on its summit. Growing as they do in so many different climates in nature, don't expect all to respond to identical conditions in the garden. Most do well with summer watering, and many even seem to need it. Most do well in a mildly acid scree, although many tolerate lime. The colder your winters, the more likely these plants are to need dry conditions. A spot below an overhang or a cloche might be needed in wet winter regions. Over 100 kinds of ice plants in 14 genera have survived Colorado winters thus far, and hardy ice plants (*Delosperma* spp.) have become a major horticultural crop internationally in a matter of decades. A few of the best are detailed here; if you have succeeded with either the yellow (*D. nubigenum*) or purple (*D. cooperi*) hardy ice plant, you may find that a palette of ice plants will grow in a sunny pocket of your garden.

Aloinopsis is largely limited in nature to the coldest parts of the Great Karoo. Paul Heiple of Denver opened up a vast arena of succulent experimentation by being the first person to plant out *A. spathulata* in a cold temperate climate. Leaves are a flattened wedge shape and of a cold, primeval gray, more suggestive of a pachyderm than a plant. The stemless flowers, an inch or so across and of the brightest baby pink, are produced any time from late February to April, depending on the site and season. They open only in bright sunlight, usually just after noon. This species requires perfect drainage and full sun and doesn't care to be waterlogged at any time of the year. Given the right spot, it persists in Colorado gardens for years, forming a massive tap root. Endemic to the higher reaches of the Roggeveld escarpment. I remember finding its leaf tips one January, barely showing in the dry, dusty, clay roadside near Sutherland; finding a colony in full bloom in October a few years later is a memory I shall never forget.

Aloinopsis malherbei is hardy only two winters out of three in Denver. But people in Salt Lake City, Albuquerque, or warmer climes will be delighted to have this astonishing plant, with coarse rosettes covered with the most amazing,

Bergeranthus jamesii, DBG.

fabric-like texture and shiny exudations of white crystals along the edges of the leaves. The flowers are large, up to 2 inches across, a deep orange sometimes tinged with pink. I ran across this once in the Roggeveld in rocky ground with only the warty tips of the leaves exposed above ground.

Aloinopsis orpenii quickly forms dense mounds of wedge-shaped leaves with stemless yellow daises in the early spring; it is very dramatic and has been hardy through wet, cold winters in Denver. The brash yellow flowers take on reddish tones as they age. The 2- to 3-inch-long, spoon-shaped leaves of *A. peersii* are attractive in their own right, possessing a fine, velvety texture; in mid- to late spring a succession of yellow-orange flowers, over an inch across, are produced in the center of the lax rosettes. This makes a very dramatic specimen for a dry crevice in a sunny garden. *Aloinopsis thudichumii* occurs over much the same range as *A. orpenii*, but the habit and flower color is much more like *A. peersii*; it is very hardy and attractive in bloom or as velvety foliage plant.

Distributed primarily at higher elevations, *Bergeranthus jamesii* has proved to be the hardiest member of a variable genus centered in the Eastern Cape. I found it on flat, sandstone pavements in the mountains above Tarkastad, where it was in bloom and seed simultaneously in March. The clumps consist of neatly tapered leaves, hard to distinguish from each other in their dense rosettes. A day-bloomer, its flowers open in the late afternoon. They are produced on pedicels 2 to 3 inches tall, reddish in bud, opening a soft orange-yellow, aging to near red.

Masses of various *Delosperma* species and hybrids in late May, DBG. The pale pink in the foreground is 'Kelaidis', a hybrid that arose in this garden and is now widely cultivated internationally.

The flowers have many petals and can approach 2 inches in diameter. This species tolerates a wide range of soils, provided they are well drained and never soggy. In Colorado, it starts to bloom in April and sports flowers until October.

A feature of many South African plants that delights the gardener is their long season of bloom, none more extended than that of *Chasmatophyllum musculinum* (jagged-leaf ice plant). This beautiful succulent from the central parts of the country produces jagged clumps of army-green leaves that eventually form a loose mat. Brassy yellow flowers start blooming in late April most years and produce repeated flushes until frost. The flowers open in the late afternoon and close early on bright mornings, but the red-stained backs of the petals are attractive even in the closed state. We have some eight collections of *C. musculinum*, each distinct. Several other members of this genus have demonstrated some hardiness, but none approach *C. musculinum* for longevity and tolerance of a wide range of soils and moisture regimes.

The yellow (*Delosperma nubigenum*) and purple (*D. cooperi*) hardy ice plants have become classics in our area over the last 15 years, treasured for their compact mats of foliage and showy flowers. A host of distinctive cousins have gradually joined them, making this genus even more desirable in gardens in the American

Delosperma sutherlandii
'Peach Star', DBG.

Delospermafloribundum,
DBG.

Delosperma cooperi
growing above 8500 feet
near Oxbow, Lesotho,
on a north-facing cliff.

West. Grown in masses and in combination with one another, they offer spectacular color for a long season of bloom and year-round texture.

Delosperma ashtonii is an abundant, tufted species of the highest mountains in South Africa. It makes a lax rosette of gray-green leaves, with 2-inch-wide, soft pink flowers with a bright white eye. It goes semi-dormant in winter, so do not despair if its appearance in spring is a bit tardy. The paler pink 'Peach Star', a European selection of the similar *D. sutherlandii*, is widely available from Jelitto Perennial Seeds.

Delosperma cooperi is widespread, especially in eastern South Africa; it grows on the Highveld, mostly on the west face of the Drakensberg, where in meadows and on fencelines between Maseru and Bloemfontein I have seen plants indistinguishable from the commonly cultivated form. It is surprising that plants that originate in this area of nearly subtropical climate can tolerate our zone 5 winters. The many high-altitude *D. cooperi* collections are much condensed in size, have a more attractive winter habit, and, most importantly, are considerably hardier. The flower color is perhaps a shade more pink. Their only drawback is that these selections do not bloom as persistently. This complex has proven to have immense value in conventional landscapes in many areas of the United States. Undoubtedly, with judicious selection and hybridization, an entire industry will arise to propagate and market this most amazing groundcover.

On a January expedition near Springfontein, Free State, I collected *Delosperma floribundum*, a compact cousin of *D. cooperi*. It produces a tuft of succulent foliage within weeks of sowing and can come to bloom within two months when sown in late spring. The first summer display may include hundreds of flowers on a single plant. The flowers are 2 inches across, shimmering lilac-pink with a large white eye that distinguishes the species easily from the better known *D. cooperi*. It is well worth growing wherever that taxon flourishes. *Delosperma floribundum* has been widely marketed by Plant Select for 15 years: it appears to

Delosperma karooicum, DBG.

RIGHT *Delosperma lavisiae* 'Lesotho Pink', DBG.

have a very wide range of climatic tolerance. It does best in sites that are not shaded or wet in the winter months and seems to be at least as winter hardy as *D. cooperi*. It is best propagated by seed, which needs warmth and light to germinate. Seed is destroyed by freezing or stratification.

Delosperma karooicum is a compact succulent limited in nature to a fairly small area of rocky knolls on Highveld steppe. It germinates promptly in warm temperatures and can come into bloom a few months after sowing. It is more diminutive than *D. floribundum*, with much smaller flowers, very starry in effect and pure white in color, barely a half-inch across. They are produced, however, from early spring to fall—seven months of showy bloom. This species makes the most impact in a tiny space, a container, or rock garden, where it can best display its dainty charms. It has been hardy, even with winter temperatures well below 0°F.

Delosperma lavisiae and *D. obtusum*—smaller, slightly more shrubby variations of *D. cooperi*—are widely distributed from the foothills to the summits of the Drakensberg. Both possess extraordinary winter hardiness but do not tolerate long periods of heat and drought in Colorado without supplemental watering. They are variable in foliage color and flower color, and several selections made by Kelly Grummons, co-owner and chief horticulturist of Timberline Gardens in Arvada, Colorado, are gaining currency in Rocky Mountain states. 'Lesotho Pink', a selection of *D. lavisiae*, is one of the hardiest of all ice plants, forming dense cushions that turn steel-gray in winter. Plants are completely smothered with flowers in spring and early summer—a spectacular ornamental.

The plant grown as *Delosperma nubigenum* for the last quarter-century undoubtedly represents a triploid or a cross with another species that has

Delosperma Fire Spinner, DBG.

LEFT *Delosperma seanii-hoganii* in a Colorado garden.

resulted in tremendous hybrid vigor. This clone should be distinguished as 'Lesotho'. The wild antecedent (which I have now seen in the Black Mountains between Sani Pass and Mokhotlong; on the cliffs below Sani Pass; and near the chain ladders on Mont-aux-Sources) is a fraction of the size (leaves a quarter-inch long, as opposed to well over an inch in the cultivated plant). The wild form has been less tolerant of garden culture, and it does not turn deep red in winter, as the widespread cultivar does. It is, however, fertile and much better suited to miniature gardens and troughs.

Delosperma Fire Spinner is a particularly flamboyant Plant Select offering. It is likely an undescribed species from alpine scree slopes in the Eastern Cape but is sometimes considered a subspecies of *D. dyeri*, which occurs some 60 miles from Fire Spinner's Sneeuberg origins. Out of bloom, the rapidly spreading mat of foliage looks much more like *D. nubigenum* 'Lesotho', although it does not take on red tints in winter. The spectacular two-toned flowers, of violet-blue and a lurid vermilion, are utterly unique. It has proved winter hardy from coast to coast in North America, although it does not bloom reliably where it does not experience a period of frost in winter.

Named in 2013, the miniature *Delosperma seanii-hoganii* is known from only two localities in the wild, occurring atop large, flat rock surfaces above 3000 meters along the Drakensberg escarpment. The foliage is a deep (almost black) purple, and the white flowers are less than a centimeter across. It is very cold hardy but needs superlative drainage and winter dryness to survive in the garden.

Hereroa is among the hardiest genera in Aizoaceae. Concentrated in the Karoo, most species, including *H. calycina*, possess a great degree of winter hardiness. They form substantial lax rosettes that can clump up to a foot across,

Boophone disticha
on Eastern Cape
karroid steppe.

Cyrtanthus epiphyticus.

with dozens of 4-cm flowers produced much of the summer. They are vespertine and challenging to photograph as a consequence. *Stomatium*, one of the largest genera in this hardy family, is mostly concentrated in the coldest parts of the Karoo, although *S. mustellinum* also occurs in the high Drakensberg. Plants are extremely variable in the form of their foliage, some being jagged. Flowers are generally 2 cm or less across—opening at dusk and resembling shaving brushes in form. Most have a very fruity, strongly sweet fragrance. *Stomatium lesliei* is especially beautifully colored—dandelion yellow.

AMARYLLIDACEAE

The amaryllis family is prized in gardens for the many species with large, often waxy flowers and elegant form. Most members of the family are bulbous and can be either winter- or summer-dormant, depending on whether they come from a winter-precipitation or summer-rainfall region. The hardiest species are generally from the latter. The foliage is often attractive—especially in some of the "tumble-weed" group. Amaryllids often have fleshy fruit that must be sown promptly (it can die if it dries out), and many take years to reach maturity. They compensate for this by being very long-lived and pest-resistant (most have high concentrations of alkaloids that are distasteful to herbivores). As a consequence, these can proliferate on overgrazed ranges, creating some spectacular displays of bloom.

One of the most widespread and variable genera is *Cyrtanthus*, which may have its center of diversity in the Eastern Cape. These all have strappy foliage and the classic six-petaled, vase-shaped flower associated with so many amaryllids. *Cyrtanthus breviflorus* is especially common from the foothills of the Drakensberg to the alpine tundra: it almost resembles a crocus in some high-altitude forms, or a daffodil without a central boss in lower-altitude forms. *Cyrtanthus epiphyticus* also has a wide range on the Drakensberg; a form collected on Naudesnek has grown outdoors in Pueblo, Colorado, for nearly two decades, often blooming heavily. Unlike their northern hemisphere cousins, many Eastern Cape cyrtanthus rebloom during the growing season if they get sufficient rain.

Two of the most dramatic genera occur in the Karoo and somewhat more sparingly on the Drakensberg proper, often producing superficially similar spherical clusters of flowers that are visible from a great distance. *Brunsvigia*, whose two large leaves hug the ground, occurs in several species with flowers of pink, scarlet, or nearly purple. The flowers of *Boophone disticha* are usually white, and its fan-like leaves are spectacular. The greater number of species in both these tumbleweed genera occur far to the west, in the much milder, winter-precipitation areas of the cape—where they are summer dormant. In the Drakensberg, these are summer-growing, blooming from November to March.

The common representative of *Haemanthus* in the Drakensberg is *H. humilis*

Agapanthus campanulatus.

ssp. *hirsutus*; it is often found on mid-altitude cliffs but can occur up to 8000 feet in elevation. Its white ball of flowers is very decorative, but the red-berried fruit that follow are even showier. Although collected high in Lesotho, this has not proved winter hardy in Colorado.

Agapanthus is a perennial endemic of southern Africa consisting of six to ten species. In the distant past, the genus was regarded as a member of the Alliaceae (onion family), although both it and the Alliaceae have been subsumed by the amaryllis family in the widest sense—the previously mentioned amaryllids are currently considered to be in the true amaryllis subfamily (Amaryllidoideae), whereas *Agapanthus* is in a parallel sister subfamily, Agapanthoideae. The flowers do somehow seem somewhat intermediate between amaryllids and onions. *Agapanthus campanulatus* is the only member of the genus that occurs widely at higher altitudes in the Drakensberg. It is fully deciduous and very cold hardy in gardens, growing continuously in Colorado for upward of 40 years. A mature clump can get over 30 inches tall if planted in rich garden soil and well watered; with dozens of flower stems, it creates a spectacular picture in the midsummer border.

ANACAMPSEROTACEAE

This is a small family (excepting *Anacampseros* and its close allies) in South Africa compared to the Americas. The flower structures across the family are surprisingly similar, but the variation in stem form and foliage shape and succulence is nothing short of astonishing in the South African xeric species. *Anacampseros rufescens* grows over a wide area and at various altitudes. Typically a plant of near desert or rocky steppe habitat in regions with relatively light frosts, it also occurs at high elevations on several ranges in the Eastern Cape. It was introduced from the summit of the Witteberg spur of the Drakensberg in 1996 and has persisted from that collection in the lists of many nurseries. The dark purplish rosettes of rubbery foliage are attractive in their own right, but the cups of refulgent pink flowers much of the summer are the real attraction—even if each blossom lasts only a few hours in late afternoon.

ASTERACEAE

The daisy family—along with the grass family—predominates in all steppe regions. Although many South African daisies superficially resemble the north temperate and South American kinds, they are often in subfamilies that don't occur outside of Africa.

Berkheya both resembles and is in fact related to north temperate thistles but differs significantly by having showy ray flowers (most Eurasian and American thistles don't). Species are extremely variable in size, but most have bright yellow flowers. Exceptions include *B. purpurea*, which has strange, crepe-like flowers in an unusual pink shade, unlike anything I can think of; it has proved amenable to cultivation in a variety of climates. *Berkheya cirsiifolia* grows in roughly the same habitats (grassy and rocky meadows, at all elevations including alpine), only it has white flowers. Both can grow over a meter tall, doing best in a porous loam that doesn't dry out deeply. They can be long-lived in gardens.

Most euryops grow as shrubs in the Western Cape—widespread in both the Fynbos and Karoo. Several South African species are popular in cultivation as ever-blooming shrubs for containers or cool conservatories; two are restricted to the highest altitudes in the Drakensberg and are completely cold hardy in continental North America. *Euryops acraeus* grows on alpine cliffs and screes in nature, bearing 2-inch bright yellow sunflowers in late spring. It is worth growing for its

Berkheya purpurea
on an Eastern Cape
alpine meadow.

LEFT *Euryops acraeus,* DBG.

foliage alone: the blue-silver leaves are extremely bright and almost succulent in texture. It makes a dense mound and can be long-lived if properly sited. *Euryops decumbens* has almost the same range, only it is found even a bit higher, on low, hummocky alpine tundra, where it forms dense cushions and mounds only 2 inches or so tall. It is studded with yellow flowers in May and June. Both are very worthy of cultivation.

Gazania species have been grown in the colder parts of the northern hemisphere for centuries as annuals, but two gazanias, albeit more delicate and smaller-flowered, have proved extremely hardy and adaptable in temperate climates. *Gazania linearis* is the hardiest, with narrow strap-shaped green leaves and 3-inch flowers that vary considerably in ray flower markings and habit throughout the Drakensberg. This self-sows abundantly—to the point of being a minor nuisance in some gardens! Most common in cultivation is *G. linearis* Colorado Gold, which resulted when plants from all over the Drakensberg were grown next to one another and produced a strain with great variability and vigor; it thrives in gardens in Vail, Colorado, above 8000 feet. *Gazania krebsiana* replaces *G. linearis* at lower elevations and in much more xeric conditions on the Karoo, where it is extremely common everywhere at all elevations. Although not as cold hardy, this plant has great utility in hot, dry regions. *Gazania krebsiana* looks like common bedding gazanias (selections of *G. rigens*) and has shown a tendency to hybridize with them.

Haplocarpha is a small genus of highly distinctive daisies that form nearly succulent rosettes of ground-hugging gray-green leaves. In *H. thunbergii*, naked

Euryops decumbens in a Colorado garden.

Gazania linearis, DBG.

stems arise producing 3-inch sunflowers at the ends that are slightly nodding in bud but which open wide and stand out dramatically for their protracted bloom period (late spring to midsummer). With soaking rains, this will continue to bloom through the summer. It will also bloom a few months after sowing seed. It is not very long-lived, so you should save seed to grow more.

The largest group of South African daisies are the everlastings (*Helichrysum* spp.), which occur across much of Eurasia as well. Hilliard and Burtt (1987) list hundreds in their treatment, which is just the southern Drakensberg in Lesotho. In Eurasia, helichrysums are plants of relatively hot, dry sites. In South Africa this genus is so widely distributed, in so many habitats, that generalizations are meaningless. The Drakensberg is likely its center of diversity, and most species from these mountains demand cooler and moister conditions than those from other parts of the country. In Colorado, South African helichrysums are best grown on granitic scree or even in peat beds. For more than ten years, we have had six species overwinter with no lasting damage. *Helichrysum* is a major impetus for exploring South Africa: dozens of choice, unintroduced species bloom throughout the growing season. Anyone seeking to make their rock garden attractive in the summer will come to love this genus.

Helichrysum milfordiae, the most popular rock garden helichrysum, has persisted in cultivation for much of the last century; it was probably collected by Helen Milford, who introduced most of the best-known South African high-mountain genera to cultivation, including *Rhodohypoxis*. This attractive plant makes dense, wooly mats on cool rock faces and adapts especially well to European rock gardens, where it is frequently seen in specialist gardens in Scandinavia, western Europe, and Great Britain.

Most helichrysums are white-flowered, but three of the loveliest come in pinks and reds. *Helichrysum adenocarpum* occurs widely around the Drakensberg, blooming in January and February—a small clump can have dozens of showy flowers. The dark, strap-shaped leaves of *H. vernum* make a neat rosette, bringing dark reddish flowers to moist meadows and forest margins in October; it is not yet in cultivation in North America. The largest of the pink helichrysums, *H. ecklonis*, has wonderfully wooly strap-shaped foliage reminiscent of lamb's ears and stems to 2 feet tall, topped with enormous flowers varying from nearly white to deep rose-red. A striking ornamental, grown well in a few gardens.

Helichrysum praecurrens is found on low, hummocky alpine tundra, where it forms dense cushions and mounds only 2 inches or so tall alongside *Euryops decumbens*. Dense dark green mats can spread to many feet across. This is positively obscured by the everlasting flowers that are over an inch across and can be bright pink as well as the commoner white. Long-lived on a north-facing crevice garden at Denver Botanic Gardens.

Helichrysum adenocarpum in sparse grass on Platberg above Harrismith, February.

Helichrysum milfordiae at the Alpine Garden, Lautaret, in the French Alps.

A 2-meter-wide mat of *Helichrysum marginatum* at Royal Botanic Garden Edinburgh, late June.

Helichrysum ecklonis at Royal Botanic Garden Edinburgh, late June.

Helichrysum praecurrens, DBG.

Helichrysum album on Sani Pass, Lesotho.

Helichrysum marginatum grows at higher altitudes throughout the Drakensberg. Plants vary in size, from 5 to 12 inches. The leaves are sparsely hairy, deep blue-green, thick, almost succulent, forming lax rosettes. A single plant can form a mat a foot or more wide with dozens of flower stems. The papery flowers, an inch or so wide, open in late June and look fresh for much of the summer. Needless to say, this is a welcome time for showy bloom. It is widely cultivated in Europe and North America. *Helichrysum album*, with similar flowers over a compact rosette, is a more strictly alpine cousin that does better in cooler conditions.

One of the most prevalent groundcovering helichrysums at high altitudes is *Helichrysum flanaganii*, a wooly-leaved mat former that can make carpets meters across. The yellow clusters of bloom lack ray flowers but are modestly decorative. This has proved extremely cold hardy and tolerant of a wide range of garden conditions in cultivation in the northern hemisphere. In dramatic contrast to *H. flanaganii*, *H. sessilioides* is a tiny cushion plant that grows on cliffs at the highest elevations throughout the Drakensberg. The papery white flower clusters are produced in mid-spring and remain in peak form for several weeks. It has grown quickly from seed and easily from cuttings. The flowers close at night, and the appearance of the plant is then so different that it is intriguing to see. This is a bona fide alpine, worthy of growing alongside the fussiest androsaces and temperamental primroses. Even more challenging, from the highest altitudes, is *H. pagophilum*, which forms vegetable sheep with time.

Widespread on subalpine and alpine rock cliffs and meadows, *Helichrysum*

Helichrysum trilineatum, DBG.

Helichrysum flanaganii at Yampa River Botanic Park, Colorado.

Helichrysum pagophilum on rocky tundra, Black Mountains, Lesotho.

trilineatum was the first shrubby species we grew at Denver Botanic Gardens, and it remains one of the most gratifying dwarf shrubs for many uses. Knowledgeable visitors assume that young specimens are a species of *Santolina*; however, *Helichrysum* tolerates more shade and moisture than lavender cotton. *Helichrysum trilineatum* can grow to a yard or more in height and even wider, although it takes a number of years to do so. The yellow, button-like heads produced in early summer are decorative for many weeks. Some lump this with *H. splendidum*, but the two species are really very distinct, albeit also very variable. *Helichrysum splendidum* has longer leaves and much larger, showier heads of bloom than *H. trilineatum*. Compact alpine races of both species are found in the Drakensberg. There is no mistaking the two in the wild, and I think both species deserve a place in most gardens. There are never enough hardy, evergreen shrubs—although strictly speaking, these are shimmering ever-silver.

Hirpicium armerioides, a striking cushion plant, has settled into cultivation in recent years; it seems to do best in a well-drained scree soil in full sun with weekly irrigation. The best form is a true, stemless alpine from the high Drakensberg, but it is found sparingly in larger forms all the way to the lower foothills. It makes a mop of hispid, deep green leaves 2 to 3 inches deep and up to a foot across in a few years. The flowers are typical, trim, white daisies that close at night to show off a neat black stripe on the reverse of the ray flowers. Flowerheads are produced in a constant succession from late spring through October.

Osteospermum Purple Mountain in a mass planting in Denver.

RIGHT *Hirpicium armerioides*.

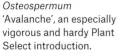

Osteospermum 'Avalanche', an especially vigorous and hardy Plant Select introduction.

Osteospermum is sometimes combined with *Dimorphotheca*, the mostly Western Cape genus of (largely) annuals. Either generic name is apt to produce attractive plants, although the most cold hardy of them, from the highest altitudes, are unquestionably *O. jucundum* and the superficially similar *O. barberiae*, which themselves are occasionally synonymized (although it seems to me that two distinct taxa are cultivated). *Osteospermum jucundum* is found in grassy meadows and screes at high elevations throughout the high Drakensberg—and usually makes small, loose mats with pink flowers in nature and the garden. *Osteospermum barberiae* is much huskier, with larger rosettes, and it is intensely aromatic: I believe this is the taxon that grows so abundantly on the Sentinel, where it makes huge mats; the flowers are a deep magenta-purple, distinct from the more pinky flowers of *O. jucundum* one finds throughout the Drakensberg proper. Both are outstanding garden plants in much of the cold temperate northern hemisphere. Their flowers obscure the foliage for months in spring and early summer, often reblooming through the whole growing season, and their evergreen foliage is attractive at all times. Plant Select has introduced three cultivars that derive mostly from *O. barberiae*—the compact Purple Mountain and the taller Lavender Mist, a pale lavender fading to white. 'Avalanche', developed in Europe, is even more vigorous and disease-free; it can form mats a yard or more across, and the flowers are almost blinding in their pure whiteness.

Othonna is found primarily in the drier parts of the Karoo and Fynbos.

Othonna capensis.

LEFT *Osteospermum barberiae,* variously colored garden seedlings.

Senecio polyodon, DBG.

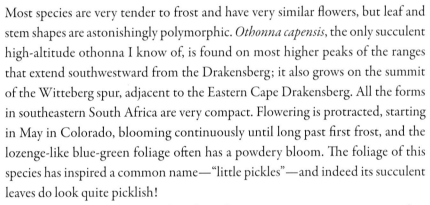

Most species are very tender to frost and have very similar flowers, but leaf and stem shapes are astonishingly polymorphic. *Othonna capensis*, the only succulent high-altitude othonna I know of, is found on most higher peaks of the ranges that extend southwestward from the Drakensberg; it also grows on the summit of the Witteberg spur, adjacent to the Eastern Cape Drakensberg. All the forms in southeastern South Africa are very compact. Flowering is protracted, starting in May in Colorado, blooming continuously until long past first frost, and the lozenge-like blue-green foliage often has a powdery bloom. The foliage of this species has inspired a common name—"little pickles"—and indeed its succulent leaves do look quite picklish!

The giant genus *Senecio* is found on all continents except Antarctica, and its members are highly variable in flower color and habit. A few of the six species in the Drakensberg with brilliant purple-red flowers are coming into cultivation. *Senecio polyodon* is the commonest of these, with small but piercing bright flowers; it grows in moist swales throughout the Drakensberg. *Senecio macrospermus* is conspicuous even out of bloom due to its massive rosette of silvery, lanceolate leaves that can stretch almost a yard in all directions, and its flowers are comparatively huge (over 3 inches across), lasting much of the summer season; this has proved amenable to cultivation in a deep, rich loam in full sun.

BORAGINACEAE

A family not nearly as prevalent in the southern hemisphere as it is in the northern—nevertheless, several striking borages are native to South Africa. *Anchusa* and *Myosotis* are highly diversified and widespread in Eurasia (*Myosotis* extends to North America as well), and both genera have representatives in South Africa. *Anchusa capensis* is found throughout the southern end of the continent. It makes an outstanding addition the garden, blooming the first year from seed and reseeding moderately in most conditions. It produces stems up to 18 inches tall (although a dwarfer race is circulating in commerce), studded with dozens of cobalt-blue, white-eyed flowers that combine with almost any other hue. Two blue-flowered South African *Myosotis* species are similar to (and apparently derived from) Eurasian species; they grow abundantly in montane meadows and forests but have not yet gained a toehold in cultivation.

BRASSICACEAE

Another family that is far vaster in the northern hemisphere—but one genus of short-lived perennials, *Heliophila*, is found throughout the Karoo and Drakensberg, in gravelly or grassy meadows. Plants bloom from late spring through summer; they are usually some shade of pink or purple, although whites are not unheard of. Their willowy stems can be almost 2 feet tall. These are ornamental and worth cultivating but are not yet in the nursery trade.

CARYOPHYLLACEAE

Most of the Caryophyllaceae in the southern hemisphere are delicate and white-flowered with the notable exception of *Dianthus*, which is common and widespread in all the major biomes—particularly in the Karoo, where a dozen species are found in white, pink, and almost yellow. I was startled to find *D. basuticus* growing in rock crevices in the Eastern Cape, looking remarkably like European *D. gratianopolitanus*! The pinks in the Drakensberg can grow 5 to 6 inches tall in the dwarfer sorts to well over a foot; most are pale pink in color.

Alpine cushion *Crassula setulosa* var. *curta* on Tiffindell, Eastern Cape.

CRASSULACEAE

The namesake genus *Crassula* is even more diverse in habit and size than *Sedum* is in the northern hemisphere (no sedums are native to South Africa). In the Drakensberg, *Crassula* species often grow in moist habitats. One of the commonest, *C. setulosa*, comes in many varieties. An especially tiny race, *C. s.* var. *curta*, was introduced to cultivation by Helen Milford in the early 1940s; it is found at the highest elevations, often on partly shady, gravelly slopes that are seasonally moist. We have found this easy to grow, provided it never dries out and is planted in part shade. *Crassula sarcocaulis* is a shrub that can grow more than a meter high, suggesting a high-altitude jade plant! It too is variable in leaf size and hardiness. It can freeze back in subzero winters but usually comes back from the base. The flower are typically a creamy white, but a deep rose-red selection is commonly grown in Europe.

Dianthus basuticus on
alpine tundra, Naudesnek.

Crassula sarcocaulis on rocky
tundra, near Mont-aux-Sources.

Cotyledon orbiculata, flat foliage
form, Eastern Cape.

Cotyledon orbiculata, cylindrical
foliage form, Eastern Cape.

Cotyledon orbiculata is found in practically every ecosystem in South Africa, from coastal cliffs to the highest alpine summits. Although germplasm from high elevations has been introduced on many occasions, seedlings rarely persist outdoors in Denver for more than a year or two before hard frost knocks them out. The foliage can be flat and paddle-shaped, or cylindrical, like a cigar. The nodding, flaring flowers are often bright red or orange, reminiscent of *Dudleya* or *Echeveria* in their comportment. Worth every effort to tame!

EUPHORBIACEAE

This enormous family of plants, another universal of steppe regions, is found on all continents but concentrated with special intensity in the deserts and semi-arid parts of South Africa, where hundreds of species occur. They can be herbaceous or shrubby—although the greatest number are simply succulent. Nondescript semi-succulent euphorbias are found throughout South Africa, but the stem succulents have much greater ornamental beauty. Three imposing species, two of them spectacular cushion-forming euphorbias, occur at higher elevations (although they have not proved as durable as the ice plants they often grow with).

Euphorbia clavarioides is very widespread on steep, sunny slopes throughout the Drakensberg, often making enormous, mounding cushions a meter or more across. It is found at very high elevations in the cold Eastern Cape mountains, whence so many ironclad hardy perennials come. Its bright yellow hypanthia turn the cushions yellow in early spring, and the seed pods, like round red peas, are very decorative in summer; but, alas, winters with prolonged subzero cold finish it off. We hope to one day develop hardy strains of this great euphorbia. Its variety *truncata*, very common in the Drakensberg and vicinity, bears an absurd

Euphorbia pulvinata on steep slopes, Eastern Cape.

Euphorbia clavarioides on rocky tundra, Tiffindell, Eastern Cape.

resemblance to the spineless hedgehog cactus localized along the border of Colorado and Utah. In the Eastern Cape this variety is often replaced by the larger and somewhat sharper *E. pulvinata*, which makes even more massive cushions. *Euphorbia pulvinata*, found here and there across much of the Karoo, has been popular as a houseplant for years, but it grows in such cold parts of South Africa that hardy forms—which would survive zone 6 gardens, at least—must exist. Doubtless this would thrive at lower elevations in cities such as Albuquerque; seed collected from plants at the highest elevations would be worth testing in the Interior West.

Farther east on the high Karoo one occasionally encounters the stem succulent *Euphorbia multiceps*, which forms a perfectly conical mound, like a tiny Christmas tree. I saw a few sizable specimens on the plateau summit of the Roggeveld escarpment—which has experienced subzero cold on rare occasions. It is much more common on the subtropical lowlands of the Western Cape.

Sebaea sedoides on moist grassland, Drakensberg foothills, KwaZulu-Natal.

RIGHT *Geranium magniflorum.*

GENTIANACEAE

This family is not nearly as widespread or diverse as it is in the northern hemisphere—or even South America. One genus, however, *Sebaea*, has many beautiful species, and one or another can be found in mesic sites, in rich soil, throughout the Drakensberg. Unlike those in the northern hemisphere, these gentians are mostly bright yellow (as in *S. sedoides*), rarely white. Several very showy tiny groundcovering sorts that grow in moist seeps have been introduced into cultivation.

GERANIACEAE

The greatest diversity in this family occurs in *Pelargonium*, which is mostly African in distribution, although two species occur in Eurasia. The handful of pelargoniums that occur in the Drakensberg are not terribly showy. *Pelargonium alchemilloides* is abundant on rocky cliffs in subalpine and montane areas, but its tiny white flowers do not inspire enthusiasm. Some spectacular species occur at slightly lower elevations and also in the Karoo, but few of these are in cultivation. There are, however, many *Geranium* species in the Drakensberg. Although not nearly as diverse as this genus is in the northern hemisphere, the South African geraniums are almost all decorative, and a few have qualities that distinguish them.

Geranium magniflorum is found primarily at the very highest altitudes in Lesotho, often growing on dry, gravelly sites. It seems to prefer a moist, loamier soil in cultivation, where it has performed very well for nearly two decades. This species has very finely divided foliage, reminiscent of a finely cut parsley, and luminous purple-pink flowers, which bloom heavily in late May and June and sporadically thereafter. Unlike northern hemisphere geraniums, which are mostly winter dormant, *G. magniflorum* is evergreen in even the coldest winters,

Rhodohypoxis baurii on Sani Pass, Lesotho.

although the foliage can be flushed with purple overtones. This superb small-scale groundcover or edging for perennial borders was promoted by Plant Select.

HYPOXIDACEAE

It is strange that members of this family (so widespread in South Africa) occur all the way to the Great Lakes region of North America and yet are absent in Eurasia. The overwhelming bulk of several hundred species in this family are restricted largely to southern Africa, where they are a conspicuous and abundant part of the vegetation—especially in the Drakensberg, where most of the genera in the family are found. *Hypoxis* species are almost always yellow-flowered and are superficially similar to one another in general morphology, although they can be radically different in size. A few enormous taxa have leaves a foot or more long and very long flower stems; other species form broad rosettes. Most look like our native *H. hirsuta*, known from even the foothills of the Rockies, although it has not been seen at its few stations in recent years. *Hypoxis hemerocallidea* has been collected from the wild on a large scale in recent decades: certain of its compounds are used to treat prostate cancer.

Rhodohypoxis baurii was introduced to cultivation by Helen Milford from the high Drakensberg, where it is extremely common. It is grown on a wide scale all over the world, even sold as a florist plant in grocery stores. The wispy rosettes of hairy foliage are nondescript until the flowers, 3 to 4 cm across, appear. A well-grown plant can produce dozens of flowers, from May through early summer, in pure white, soft pink, deep rose-purple-pink, and every shade in between. It comes from habitats that are often moist in the summer months but frozen and very dry in the winter—conditions that are not easily reproduced in most

Gladiolus saundersii at Laporte Avenue Nursery, Fort Collins, Colorado.

Gladiolus ecklonii on alpine grassland above Sentinel, Free State.

gardens. As a consequence, gardeners often resort to growing these in containers and protecting them from excessive frost or wetness in winter by storing them in refrigerators or alpine houses. I have seen large colonies of these grown outdoors in Sweden (where they are mulched in winter to repel excess moisture) and also in England and the United States, where clever gardeners have developed methods for growing them in the ground with some sort of overhead protection. An aggressive program of testing might yield some reliably hardy forms of this uniquely decorative South African wildflower.

IRIDACEAE

The namesake genus *Iris* is entirely northern hemisphere in distribution, but the bulk of genera in the family are overwhelmingly concentrated in southern Africa. The greatest center of diversity is the Western Cape Fynbos and Karoo, with a secondary center of endemism in and around the Drakensberg, whence many of the showiest and most adaptable species come. *Crocosmia*, the other stalwart South African irid, is a mainstay in north temperate gardens; most species are either woodland plants (*C. aurea*) or restricted to streamsides and very moist habitats (*C. paniculata*)—hardly characteristic of the more exposed steppe habitats where many other irids thrive. Crocosmias in cultivation are invariably vigorous hybrids better adapted to garden conditions.

A great many *Gladiolus* species are found only in the Drakensberg. *Gladiolus flanaganii* (suicide lily) is largely restricted to dangerous cliffs there, where zealous enthusiasts have plunged in its pursuit (hence the common name); introduced in the 1990s, it has proved amenable to cultivation. The showy *G. dalenii* is the best-known hardy gladiolus. It is widespread in Africa, growing in grassland at all altitudes below the alpine from the Cape provinces all the way to Ethiopia; it seems to be uniform in the Drakensberg—large flowers are heavily patterned with dark speckles. *Gladiolus saundersii* looks somewhat similar from a distance, but the piercing coral-red flower lacks stippling and has a wonderfully contrastful white throat; this species, which generally grows at the highest altitudes in Lesotho and the Drakensberg, has come back reliably for years in Denver. *Gladiolus ecklonii* is more strictly alpine than the preceding, and much smaller; it often comes in a dark coloration.

For many years now, a tall, yellow-flowered iris look-alike has been making the rounds of iris enthusiasts throughout the United States. Superficially, it looks like a bright yellow Siberian iris, but each flower persists only a day, and the long, fibrous-based leaves, which splay widely, are evergreen, indicating that this is no simple iris. Indeed, it's not an iris at all, but *Moraea huttonii*. This is the first moraea most of us have encountered. Several dozen moraeas occur at lofty altitudes in the Drakensberg, and more occur in other high ranges throughout South Africa,

ABOVE *Gladiolus dalenii*, Eastern Cape.

LEFT *Moraea huttonii*, DBG.

BELOW Comparatively huge flowers of *Moraea alticola*, Eastern Cape Drakensberg.

Tritonia drakensbergensis,
Tiffindell, Eastern Cape.

Mentha longifolia, DBG.

in many colors and habits. Dozens of brilliant *Moraea* species in the Western Cape demand dry summers and mild winters; the showy Eastern Cape species, however, like rich loam and regular moisture. All resemble iris, but *Moraea* has a shimmering, evanescent beauty distinct from *Iris*. And many moraeas bloom much later in the season. Alas, *M. huttonii* is almost the only high-altitude species commonly grown. Another—the much larger, ivory-and-yellow *M. alticola*—is getting around as well and is starting to appear in some catalogs.

Tritonia drakensbergensis, a miniature cousin of *Gladiolus* from the Drakensberg, carries its peachy or pink flowers on short stems in summer. It has established a toehold in cultivation and should prove a useful bulb for temperate gardens. *Tritonia disticha* is similar but usually larger.

LAMIACEAE

The mint family is widespread in South Africa, although perhaps it doesn't figure as largely as it does in Eurasia and parts of the American Southwest, where mints are a major element in the flora. Several South African *Salvia* species are showy—but these don't occur in the colder steppe. In *Plectranthus*, only one or two herbaceous species grow fairly high in the hills, and some unusual mints are very striking in the Little Berg (the foothills of the Drakensberg in KwaZulu-Natal), such as *Syncolostemon macranthus*, which makes a fine show in mid-altitude meadows but has yet to show up in north temperate gardens. *Mentha longifolia* is the only true mint in South Africa and, like a few other South African taxa in several families, shows up again in the Himalaya. This is possibly the most attractive mint, with showy pink clusters of bloom and wonderful silvery foliage. Not as invasive as its northern hemisphere cousins, it does spread quickly at the root. It seems to be more drought tolerant than north temperate mints but needs some irrigation in semi-arid gardens.

MALVACEAE

Recent treatments have sunk the chocolate family, Sterculiaceae, into the much more widespread Malvaceae. Members of this former group occur across much of southern Africa, with a great many found in the high Karoo and the Drakensberg. Their flowers remind me of nodding oxalis rather than mallows, although the foliage is perhaps more like the latter. One tiny scarlet-flowered species introduced from the top of Hantamsberg, *Hermannia stricta*, has become a mainstay of rock gardening in England over the last 20 years, often shown and grown in pots. It makes a dense clump 5 to 8 inches tall and broad with dozens of nodding vermilion parasol flowers an inch or two across for much of the growing season. Many more potentially hardy *Hermannia* species have yet to be introduced to cultivation.

Merxmuellera macowanii
near Balloch in the Eastern
Cape mountains.

POACEAE

As in the other major steppe regions, the grass family throughout southern Africa constitutes the greatest volume of plant matter, and a large proportion of the biodiversity herein. A surprising number of southern African grasses are closely allied to those in the northern hemisphere: *Aristida*, *Eragrostis*, *Festuca*, and *Panicum* are shared with most other steppe regions. But Africa also has lots of endemic grasses, many of them extremely important economic plants. What follows is a mere sampling.

Africa is the homeland of all *Cynodon* ("dogtooth") grasses, including *C. dactylon* (Bermuda grass), although they are now grown in subtropical regions worldwide for turf and grazing. In many climates they are regarded as wildly invasive, but in the harsh plains of Colorado a miniature *C. dactylon* was collected in

Harpochloa falx, DBG.

the early 1980s from a farmstead, where it had likely evolved to tolerate the more extreme conditions by "hunkering down." This race is now being marketed and sold widely in western America by High Country Gardens.

Harpochloa falx, a clump-forming native of mid-elevation grasslands in the Drakensberg, makes a striking specimen in the front of the border, with silvery foliage a foot or so tall, and flowers held at roughly 18 inches. They uncannily mimic the eyelash flowers and seedheads of blue grama (although the two grasses are not very closely related).

Merxmuellera macowanii is a finely textured bunchgrass widely distributed at higher altitudes throughout the Drakensberg, often growing with *Themeda triandra*. On deep rich soils, it often makes clumps a meter tall and a meter or more across. Not yet widely grown in cultivation in the northern hemisphere.

Themeda triandra is a major constituent of high-altitude grassland in the Drakensberg and neighboring mountains; it also grows in East Asia, and in South Africa, it constitutes a large proportion of what is called Sourveld, to distinguish it from the "sweeter" grasses that grow on alkaline substrates at lower elevations. Its substantial seedheads grow on stems that can exceed 2 feet in height; they are striking from the time the plant comes into bloom in early summer through much of the successive year. The graceful sweep of tawny green grass blades is welcome in the garden, even in winter when they turn a deep auburn.

Alchemilla colura on
Sani Pass, Lesotho.

ROSACEAE

The rose family is conspicuous in South Africa for its paucity: the only wide-spread genus of woody roses in the region is the tiny-flowered *Cliffortia*, which is found primarily in the Cape Floristic Province but also in the Drakensberg. Interestingly, several species of *Alchemilla* (lady's mantle) are found in the region. *Alchemilla colura*, common in the Drakensberg in moist areas, is superficially similar to other members of the genus. I don't believe any of the South African alchemillas are in general cultivation. *Agrimonia eupatoria*, which occurs in South Africa, is also found in Eurasia and North America.

SCROPHULARIACEAE

The figwort family has undergone some major realignment, as genetic studies reveal new relationships that are not apparent to the naked eye. Many former figworts are now in Plantaginaceae. Wherever they are lumped, there is no question that plants in these groups are a treasure trove for gardeners, and a welter of genera stretch from the higher reaches of the Karoo and Drakensberg.

Diascia is one of relatively few South African genera that have gained wide currency with gardeners. Several showy annuals are restricted to the Western Cape; most perennial species grow in and around the Drakensberg. One in particular deserves to be singled out: *D. integerrima* possesses a great deal of vigor and tolerance of extremes of heat and cold. The typical wild form grows a yard or more tall and flops. I remember Kelly Grummons of Timberline Gardens coming by in the early 1990s, admiring the large clump housed up against what was then the

Diascia integerrima Coral Canyon at Mesa Xeriscape Demonstration Garden, Colorado Springs.

Selago flanaganii with helichrysums near Oxbow, Lesotho.

Zaluzianskya ovata in a Colorado garden.

DBG Alpine House; he remarked, "If we only had a compact form of this, it would be a winner." A year or so later I found a compact colony near Rhodes in the Eastern Cape. A pinch of seed from these resulted in Coral Canyon, a landscape plant that loves to grow in full sun and is reliably perennial in zone 5, given a modicum of good soil and water. Plant any one of the hundreds of *diascia*s now in cultivation in a typical garden bed; most will melt away before midsummer, and few indeed will come back a second year. Coral Canyon plants have persisted for a decade or more in an optimal spot. And I have seen it (ever so sparingly) growing around the country. The simple, sad fact is that this best of class is still essentially unknown by the great majority of keen gardeners in America: more's the pity!

Selago is sometimes placed in its own family. Bright lavender-blue spires much of the summer over trim mounds of foliage suggest *Veronica*, and many striking species in the Drakensberg occur at all altitudes, mostly on rocky, meadow habitats. *Selago flanaganii* is one of the most widespread and beautiful, although it has not persisted for us in the garden. The more capitate and clumping *S. galpinii*, restricted to the highest Eastern Cape subalpine zone, often has deep violet-blue flowers—but is not yet in cultivation.

Western Cape *Zaluzianskya* species were grown as annuals for much of the last century, but the perennial Drakensberg species were only lately introduced. One of the loveliest of these is undoubtedly *Z. ovata*, which generally grows on cliffs and rocky sites at higher elevations throughout the Drakensberg.

Aloe aristata and *Haemanthus humilis* ssp. *hirsutus* in the Eastern Cape.

XANTHORRHOEACEAE

Aloes are thought of as subtropical or tropical in their distribution, and indeed the great bulk of the genus is found in tropical and subtropical southern Africa, thinning out northward on the continent (and making it a short way into Asia). Although aloes generally grow in areas with little or no frost, a surprising number grow at high elevations in the Karoo and Drakensberg, in areas with frequent heavy winter frost. These are primarily in the "grass aloe" section of the genus, which includes *Aloe ecklonis*—the most common of this section in the Drakensberg. They name grass aloes (on account of their grassy foliage, which dies down in winter) is all the more appropriate, since they invariably seem to grow in sparse grassland.

Aloe ecklonis usually occurs in sparse meadows and rocky habitats, in striking forms with bright orange or nearly scarlet flowers. A lemon-yellow form prevails on Platberg north of Harrismith. This has overwintered several times for us at Denver Botanic Gardens (and is easy to grow in warmer areas like California), although none have persisted long enough to develop reliably hardy strains in the colder, continental parts of the northern hemisphere. The compact habit and herbal uses should propel its utility in gardens—if only a hardy strain could be bred!

Aloe aristata is one of the more common aloes throughout the higher Karoo

Kniphofia caulescens,
Oxbow, Lesotho.

of the Eastern Cape; it is also found in the Drakensberg up to nearly alpine heights. A genotype originating near Semonkong, Lesotho, survived at Timberline Gardens in Arvada, Colorado, for one winter but has not done so again. Nor has *A. polyphylla*, despite its impeccable high-altitude credentials, yet come through a Colorado winter, although many plants have been sacrificed. This spiral aloe has become distressingly rare in Lesotho but is well entrenched in cultivation—especially in northern California.

The failure of *Aloe* in colder steppe climates is more than offset by the marked success and longevity of most *Kniphofia* species that have been tested here. Throughout the 20th century, kniphofias were among the very few South African plant commonly grown on North America's Great Plains, where they were almost always labeled *K. uvaria*, a blatant misnomer: they invariably represented hybrids involving *K. baurii* and other species from the eastern parts of South Africa—moderately attractive, certainly, but hardly tapping into the enormous diversity of this genus. Almost all cultivars offered by the U.S. nursery industry are hybrids bred in England or Holland for many traits—adaptability to a steppe climate *not* being one of them. Gardeners who grow species kniphofias

Kniphofia caulescens
in an alpine meadow,
Tiffindell, Eastern Cape.

are often surprised by their vigor and performance in comparison to these often tender or delicate hybrids. Chief among these has to be *K. caulescens*.

Kniphofia caulescens is an enormously variable and spectacular high-altitude wildflower, from the highest elevations throughout most of Lesotho and neighboring KwaZulu-Natal and Eastern Cape provinces. Its foliage is frequently bright silvery blue—almost resembling a yucca from a distance. The flower stems are stalwart and tall (often over a meter high), with vivid scarlet and white or yellow flowers in late summer and fall. It often occurs in vast numbers in fairly wet, marshy sites in nature; nevertheless, it does not seem to like to be too wet in cultivation and in fact has a wide tolerance for different soils and irrigation regimes in gardens.

Kniphofia northiae (occasionally growing near *K. caulescens*, although blooming much earlier) is perhaps the most distinctive and dramatic foliage plant in the genus. It must be placed with care in your garden, for this sumptuous plant can form rosettes 5 feet across. The shape of a cluster of plants looks much more

Kniphofia ichopensis.

like a lax agave than a torch lily, perhaps even more like a gargantuan tillandsia. The flowers quickly dispel this illusion: they form a huge cone of soft orange and yellow tubular flowers 2 feet long or more on mature specimens. This species appears to adapt readily to a wide range of garden conditions—although in nature it seemed to always grow at the highest elevations, on steep, wet slopes, often next to running water. It must be one of the first of the genus to bloom in spring, since the flowers were long past in early January in the wild—they bloom much of May in the northern hemisphere. Few plants make such a dramatic statement in the garden.

Kniphofia hirsuta, from subalpine grassy steppe, has gained a toehold in gardens since it was introduced to cultivation in the early 1990s. It is one of the smallest species, often barely a foot tall in bloom. It makes a trim clump with silvery foliage that is minutely hairy (unique in the genus). Many more species have been cultivated in recent decades by enthusiasts, each of which seems to possess some quality or merit that justifies a spot in a plantsman's garden!

Kniphofia triangularis is one of the few species commonly offered by U.S. nurseries. The form in cultivation—with trim cones of brilliant tangerine flowers—looks most like the mid-elevation forms from the Little Berg. On Naudesnek, the brilliant, bi-colored flowers of the type open a brilliant coral-red, aging to a cooler yellow. The very thin grassy foliage is attractive in the garden, and by late summer, a mature clump with dozens of flowers makes a stunning focal point. *Kniphofia thodei* resembles a red and white version of *K. triangularis*. Coming as it does from the highest ridges of Lesotho, it should prove very hardy in gardens; it is rarely cultivated.

Kniphofia stricta is a wonderful, long-blooming, mid-sized torch lily in the garden. Lower-elevation forms can have much brighter cones of hot orange, although the typical forms are less brightly colored, with dull green and yellow as well as orange in the flower. It has proven to be one of the most drought-tolerant species, well worth the garden space.

Here and there throughout the eastern foothills of the Drakensberg you are very likely to find variable representatives of a group of species with some of the most distinctive flower forms in the genus. Rather than forming the characteristic cone-shaped inflorescence, the species allied to *Kniphofia laxiflora*, *K. ichopensis*, and *K. rufa* make very lax heads with flowers clearly separated from one another—somehow suggestive of some aloe blossoms. These vary tremendously in size, habit, and color—cool greenish forms are common, as are more typical yellows and oranges. *Kniphofia rufa* is a mysterious entity that can occur in a bright orange form on the Little Berg south of Underberg, growing on a cool slope with *Helichrysum adenocarpum* and bright purple *Dierama latifolium*. One of the most amazing kniphofias I have ever seen is the pure white- and coral-flowered form of *K. angustifolia*, possibly a hybrid with *K. ichopensis*, found by Ernie and Marietta O'Byrne on a January hike above Njesuthi in the KwaZulu-Natal Drakensberg. A superb form of *K. ichopensis*, selected and propagated by Ellen Hornig of Seneca Hill Perennials, is a clear yellow, quite dwarf, and very hardy. Another strange kniphofia I encountered not far from the chain ladders above Sentinel was *K. ritualis*. This forms very broad-leaved, shiny rosettes, with a sparser than normal head of orange flowers that fade to a dull chartreuse.

Kniphofia porphyrantha forms very dense, low tufts of foliage, with flowers on stems often less than a foot high. They seemed to be uniformly bright, primrose-yellow on Sentinel. The heads were also very squat compared to other species. A much longer-flowered, more crimson-colored plant growing in the woodland chasms leading to the summit of Platberg, near Harrismith, was also identified as *K. porphyrantha*. These two forms are utterly distinct to my eyes—perhaps it is a variable taxon, or it changes dramatically as it blooms.

Another kniphofia that is very unlike any in cultivation is *Kniphofia acraea*. This grows a hundred or more miles south of the Drakensberg on mountain summits north and slightly west of Cradock in the Eastern Cape. Here, at nearly 8000 feet, the meadows look like the High Drakensberg, and winter temperatures are bitterly cold for months on end. At summer's end, a medium-sized kniphofia fills the meadow with cool, ivory-colored pokers.

Possibly the most peculiar species in the genus is *Kniphofia typhoides*, with long, somewhat tapering flower stems. In this species the flowers are dark purple, nearly black. Any of the larger kniphofias would certainly make a marvelous specimen plant in a perennial border or other strategic spot in the home landscape, wild garden, or larger rock garden. They provide colors and textures unimaginable in northern hemisphere perennials.

There are at least two really homely kniphofias. On Naudesnek and the lesser known Bastervoetpad, you may see a skinny, pale, chartreuse-colored thingum a foot or two tall with a crook in its neck. You may finally pause long enough to examine one and realize that (my heavens!), it really is a kniphofia! This is *Kniphofia parviflora*, the ugly duckling of the genus. Or perhaps that honor may go to *K. brachystachya*? There may be other forms of this, but once, near Katberg, I found a colony of these with flowers the precise shade of the mud wherein they grew.

Kniphofia porphyrantha in alpine steppe, Amatola Mountains, Eastern Cape.

BIBLIOGRAPHY

Aguiar, Martin R., and Osvaldo E. Sala. 1998. Interactions among grasses, shrubs, and herbivores in Patagonian grass-shrub steppes. *Ecología Austral* 8:201–210.

Arnold, T. H, and B. C. de Wet. 1993. *Plants of Southern Africa*. NBI.

Barr, Claude. 1986. *Jewels of the Plains*. University of Minnesota Press.

Bean, Anne, and Amida Johns. 2005. *Stellenbosch to Hermanus South African Wild Flower Guide* 5. Botanical Society of South Africa.

Birks, John. 2006. *Grupo Erskine-Central Patagonia*. eecrg.uib.no/projects/AGS_BotanyExp/Central-Patagonia/2006PatagoniaExpedition.pdf.

Bishop, Ellen Morris. 2006. *In Search of Ancient Oregon*. Timber Press.

Cather, Willa. 1918. *My Antonia*. Houghton Mifflin.

Correa, Maevia N., et al. 1969–98. *Flora Patagonica*. 7 vols. Instituto Nacional de Tecnologia Agropecuaría.

Cowling, Richard, and Dave Richardson.1995. *Fynbos*. Fernwood Press.

Cronquist, Arthur, et al. 1972. *Intermountain Flora* 1. Hafner Publishing.

Cushman, Ruth Carol, and Stephen R. Jones. 1988. *The Shortgrass Prairie*. Pruett Publishing Co.

Del Valle, Hector. 1998. Patagonian soils. *Ecología Austral* 8:103–123.

de Dios, Julián. 2009. *Patagonia Complete Guide*. de Dios Editores.

Doutt, Richard L. 1994. *Cape Bulbs*. Timber Press.

Germishuizen, Gerrit. 1997. *Wild Flowers of Northern South Africa*. Fernwood Press.

Gledhill, Eily. 1981. *Eastern Cape Veld Flowers*. Creda Press.

Goldblatt, Peter, ed. 1993. *Biological Relationships between Africa and South America*. Yale University Press.

Goldblatt, Peter, and Pauline Bond.1984. Plants of the Cape flora. *Journal of South African Botany* supplementary vol. 13. Botanical Society of South Africa.

Goldblatt, Peter, and John Manning. 1998. *Gladiolus in Southern Africa*. Fernwood Press.

Goldblatt, Peter, et al. 2004. *Crocosmia and Chasmanthe*. Timber Press.

Green, Lorraine, and Marcela Ferreyra. 2011. *Flores de la Estepa*. Zagier & Urruty.

Hardy, Betty. 1988. *Drakensberg Flowers*. Berg Books.

Hilliard, O. M., and B. L. Burtt. 1987. *Botany of the Southern Natal Drakensberg*. National Botanic Gardens.

Joffe, Pitta. 1993. *The Gardeners' Guide to South African Plants*. Delos.

Lauenroth, William K. 1998. Guanacos, spiny shrubs and the evolutionary history of grassing in the Patagonian steppe. *Ecología Austral* 8:211–215.

Lauenroth, William K., and Ingrid C. Burke, eds. 2008. *Ecology of the Shortgrass Steppe*. Oxford University Press.

León, R., et al. 1998. Grandes unidades de vegetación de la Patagonia extra andina. *Ecología Austral* 8:125–144.

Leopold, Aldo. 1949. *A Sand County Almanac*. Oxford University Press.

Le Roux, Nanalise, and Ted Schelpe. 1997. *Namaqualand South African Wildflower Guide*. Botanical Society of South Africa.

Manning, John. 2001. *Eastern Cape South African Wildflower Guide* 11. Botanical Society of South Africa.

———. 2007. *Field Guide to Fynbos*. Struik.

Manning, John, and Peter Goldblatt. 1996. *West Coast South African Wild Flower Guide* 7. Botanical Society of South Africa.

Manning, John, et al. 2002. *The Color Encyclopaedia of Cape Bulbs*. Timber Press.

McEwan, Colin, et al., eds. 1998. *Patagonia*. Princeton University Press.

McGrath, Jim. 2010. *Flora of the Uplands and Adjacent Canadian River Floodplain*. New Mexico State University.

McGregor, Ronald L., et al., eds. 1986. *Flora of the Great Plains*. University Press of Kansas.

McNab, W. Henry, and Peter E. Avers. 1994. *Ecological Subregions of the United States*. www.fs.fed.us/land/pubs/ecoregions/index.html.

Meinig, D. W. 1993. *The Shaping of America*. Yale University Press.

Mucina, Ladislav, and Michael C. Rutherford, eds. 2006. *The Vegetation of South Africa, Lesotho and Swaziland*. South African National Biodiversity Institute.

Paruelo, José M., et al. 1998. The climate of Patagonia. *Ecología Austral* 8:85–101.

Peterson, Pam, ed. 2001. *Biological Soil Crusts*. U.S. Dept. of the Interior, BLM, National Science and Technology Center, Information and Communications Group. Vers. Technical Reference 1730–2. soilcrust.org/crust.pdf.

Pooley, Elsa. 1998. *A Field Guide to Wildflowers KwaZulu-Natal and the Eastern Region*. Flora and Fauna Publications Trust.

———. 2003. *Mountain Flowers*. Flora and Fauna Publications Trust.

Province of British Columbia. 2013. *Ecoregion Unit Descriptions*. env.gov.bc.ca/ecology/ecoregions/dryeco.html.

Ridley, Glynis. 2010. *The Discovery of Jeanne Baret*. Crown Publishers.

Schmitz, Marthe. 1982. *Wild Flowers of Lesotho*. ESSA Ltd.

Schumann, Dolf, et al. 1992. *Ericas of South Africa*. Fernwood Press.

Secor, R. J. 2009. *High Sierra*. Mountaineers Books.

Sheader, Martin. 2013. *Flowers of the Patagonian Mountains*. Alpine Garden Society.

Smith, Gideon, et al. 1998. *Mesembs of the World*. Briza Publications,

National Botanic Gardens, NBG Kirstenbosch.

Spence, John R. 1999. *Climate of the Central Colorado Plateau, Utah, and Arizona.* sbsc.wr.usgs.gov/cprs/news_info/meetings/biennial/proceedings/1999/pdf/15_Spence.pdf.

St. John, Alan D. 2007. *Oregon's Dry Side.* Timber Press.

Takhtajan, Armen. 1986. *Floristic Regions of the World.* University of California Press.

Tarleton State University. n.d. *Sagebrush Shrub Steppe.* tarleton.edu/Departments/range/Grasslands/Shrub%20Steppe/Shrub%20Steppe.htm.

Taylor, Ronald J. 1992. *Sagebrush Country.* Mountain Press Publishing.

Thorne, Robert F. 2008. *Phytogeography of North America, North of Mexico.* floranorthamerica.org/Volume/V01/Chapter06.

Tidestrom, Ivar. 1925. *Flora of Nevada and Utah.* Washington Government Printing Office, Smithsonian Institution.

Tiehm, Arnold. 1996. *Nevada Vascular Plant Types and Their Collectors.* New York Botanical Garden.

Trauseld, W. R. 1969. *Wild Flowers of the Natal Drakensberg.* Purnell.

Trimble, Stephen. 1989. *The Sagebrush Ocean.* University of Nevada Press.

Trowbridge, Wendy, et al. 2013. Explaining patterns of species dominance in the shrub steppe systems of the Junggar Basin (China) and Great Basin (USA). *Journal of Arid Land* 5(4):415–427.

van der Walt, J. J. A. 1988. *Pelargoniums of Southern Africa.*

van Dijk, Hanneke. 2004. *Agapanthus for Gardeners.* Timber Press.

van Rooyen, Gretel, and Hester Steyn. 2004. *Cederberg South African Wildflower Guide* 10. Botanical Society of South Africa.

Vega, Santiago G. de la. 2006. *Patagonia.* Contacto Silvestre Ediciones.

Weber, William. 2003. The Middle Asian element in the southern Rocky Mountain flora of the western United States. *Journal of Biogeography* 30:649–685.

Weber, William, and Ronald C. Wittmann. 2012. *Colorado Flora: Eastern Slope.* 4th ed. University Press of Colorado.

West, Elliott. 1998. *The Contested Plains.* University Press of Kansas.

Whiteman, C. David. 2000. *Mountain Meteorology.* Oxford University Press.

Zappettini, Eduardo O., and Jose Medía. 2009. The first geological map of Patagonia. *Revista de la Asociación Geológica Argentina* 64:55–59.

USEFUL CONVERSIONS

INCHES	CM
¼	0.6
½	1.3
1	2.5
2	5.1
3	7.6
4	10
5	13
6	15
7	18
8	20
9	23
10	25
20	51
30	76
40	100
50	130
60	150
70	180
80	200
90	230
100	250

FEET	METERS
1	0.3
2	0.6
3	0.9
4	1.2
5	1.5
6	1.8
7	2.1
8	2.4
9	2.7
10	3
20	6
30	9
40	12
50	15
100	30
200	60
300	90
400	120
500	150
600	180
700	210
800	240
900	270
1000	300
2000	610
3000	910
4000	1200
5000	1500
6000	1800
7000	2100
8000	2400
9000	2700
10,000	3000

TEMPERATURES

$$°C = 0.55 \times (°F - 32)$$
$$°F = (1.8 \times °C) + 32$$

PHOTO CREDITS

INDEX

Plant Primer entries appear in bold type.

ABOUT THE AUTHORS

MICHAEL BONE, CURATOR OF STEPPE COLLECTION

Michael Bone works in plant propagation and production and is the curator of the steppe collection at the Denver Botanic Gardens, where he has focused his work on seed collection and the study of steppe plants and ecology. Most of his fieldwork has been in western North America, but he has traveled to Central Asia to study plants from the steppes and mountains there. He is also actively involved in the Plant Select program and the International Plant Propagators' Society.

DAN JOHNSON, CURATOR OF NATIVE PLANTS,
ASSOCIATE DIRECTOR OF HORTICULTURE

Dan Johnson has been gardening for as long as he can remember. He travels throughout the West and Southwest in search of unusual and underused native plants for trial in Colorado's semi-arid steppe climate. His forays to similar regions of the world help further broaden the palette of plants suitable for western gardens. He has created and maintains some of Denver Botanic Gardens' most beautiful and self-sustaining native and xeric gardens.

PANAYOTI KELAIDIS, SENIOR CURATOR AND DIRECTOR OF OUTREACH

Panayoti Kelaidis represents Denver Botanic Gardens in educational, professional, and promotional endeavors as an expert in horticulture, science, and art. A past president of the Rocky Mountain Chapter of the North American Rock Garden Society (NARGS) and the American Penstemon Society, he is the recipient of the Award of Excellence from National Garden Clubs and the 2000 Arthur Hoyt Scott Medal from Swarthmore College.

MIKE KINTGEN, CURATOR OF ALPINE COLLECTIONS

Mike Kintgen oversees the Alpine Collection and nine gardens at DBG including the Rock Alpine Garden and South African Plaza. He interned at both the Chicago Botanic Garden and the Rhododendron Species Foundation. He has been a member of NARGS since 1993 and is past president of its Rocky Mountain Chapter. He lectures internationally on Denver Botanic Gardens and its focus on semi-arid steppe and high-elevation floras of the world.

LARRY G. VICKERMAN, DIRECTOR
DENVER BOTANIC GARDENS AT CHATFIELD

Larry Vickerman obtained a B.S. from Colorado State University in landscape management and an M.S. in not-for-profit management from University of Washington. He currently directs Denver Botanic Gardens at Chatfield, a 700-acre farm and public garden in Littleton, Colorado. He has worked in public horticulture and landscape restoration since the late 1980s, and he also maintains an active role in the family ranching business.